NutriWine

By

Ralph Quinlan Forde

Sláinte!

Ralph

Published in 2011 by Health E Books Limited

First Edition

British Library C.I.P.

A CIP catalogue record for this title is available from the British Library.

ISBN: 978-0-9571318-5-9

Disclaimer

The publisher and author make it clear that this work is for general use and may not be substituted for medical advice by a qualified medical professional. Nothing in this book should be construed as an attempt to offer medical advice, opinion or engage in the practice of medicine. The author and publisher shall have neither liability nor responsibility to any person or entity with respect to any loss or damages arising from the information contained in this book.

Sales enquiries please contact rqf@nutriwine.net

To

Professor Leo Pyle

Ian Scott Naomi Mathew

Isco

Javier & Checho

Contents

Preface

Champagne?! You treat depressives with Champagne?...Wow.

Those were the words of the high end magazine editor in response to my answers in an interview born out of a story beginning to circulate about my clinical work. Yes indeed I do treat depressives with Champagne. You see, wine has a very long tradition in preventative medicine world wide and particularly in France. A fashion photographer who also worked for the magazine came to consult with me. As part of the healing protocol for them I included a glass of Champagne a day for two weeks to lift my patients' spirits.

When the magazine found out about this they ran straight to the phone.

So welcome to the world of wine - a $107 billion industry. The USA is now the biggest consumer of wine today overtaking France and Italy. The global wine industry will soon produce over 27 billion litres of wine a year. Wine enthusiasts in the USA are some 80 million people. Thomas Jefferson, American President who was a wine enthusiast and the visionary founder of the republic, would be very proud indeed.

French physicians actually prescribe wine for different ailments as part of a preventative medicine protocol with their patients. This long and ancient tradition of wine as a preventative medicine should not be lost which is why I am writing this book. We need to re-engage western medicine with wine and its health benefits. Wine has been a big part of the global materia medica for millennia.

Medicine without wine has actually cut itself off from its ancestral lineage of integrative and preventative medicine. Prohibition caused the impasse between wine and health but wine has always been a culture of moderation and studies demonstrate this. A whole generation of physicians have been educated without ever having been taught the health benefits of moderate wine drinking. With one out of eight Americans soon to develop diabetes we need to be open to all solutions, even natural ones like wine, to prevent this epidemic.

The French Paradox is taken up with why French people eat more fat daily than Americans but suffer less heart disease in a modern world. They are also slimmer. Much slimmer. The French are not haunted by obesity the same way as in the USA. Medical research points to the effect wine has on the French in keeping them healthy and they live longer too. Wine contains some powerful biochemicals such as vitamins, minerals, amino acids and especially antioxidants that can boost peoples' health. These nutrients are in a form that nature designed and not what marketeers have invented. Wine contains natural vitamins not chemical ones. Some people even say wine is close to the composition of blood in ways.

We need to reconnect people to the blood of the earth once again. Some of the healthiest longest living cultures of the world regularly drink wine in their diets. There is no doubt now that diet plays a crucial role in health according to scientific studies. If we add wine to that healthy diet regimen you can immediately help boost your health and wellbeing in terms of the amount of antioxidants you ingest and fat managing properties you add. Wine has been shown to help heart disease, cancer, obesity and even offset dementia. These are just some of the ways that wine appreciation can contribute to your wellbeing, prevent a heart attack and even extend your lifespan. However more research needs to be done which will reveal more astounding health factors if the example of resveratrol is anything to go by as a 'magic bullet'.

The wine industry is awash with lots of great characters and fascinating innovators. Wine is far more than just Chateau, taste, bouquet and price scale. Wine appreciation is a way of life that generates social communities and new experiences. Wine is an art form imbued throughout with an alchemy from the earth and the sun. This alchemy is imparted to us through the medium of wine. This same alchemy can heal us and the ancients in their universal wisdom have long known this. Modern people are described as suffering a form of starvation that is called 'Dionysian' that wine is said to cure. This starvation refers not just to food hunger but a hunger for the joy of life itself which is very prevalent in a society today that lives to work rather than works to live.

China is fast becoming a producer of wine and Hong Kong has now become the centre of fine wine auctioning surpassing New York. How we influence the global wine industry as the people who buy wine, such as opting for organic green wine, can hugely influence how they develop. In doing so we can also help sustain the global ecosystem and inspire a green movement within China with the new young generation of wine makers on their way.

Wine is also under threat from climate change and the industry could entirely disappear within 20 years if we don't reduce levels of greenhouse gases in our environment. Due to global warming, alcohol levels are increasing which is changing the traditional taste of wines. Vineyards are being affected and some even destroyed in flooding and fires from the instability of the weather systems due to climate change. We don't need to wait that long to see this outcome. Australian vineyards have already suffered terribly from climate instability in recent years. The good news is we now have the first carbon zero winery and a bodega in Spain is heated using geothermal power from the earth. So it can be done.

The industry also needs to stop looking in the mirror and start looking out the window. For far too long the industry has been a bit stagnant and not directly engaging with consumers.

There are now over 14 million conversations about wine taking place online every year in social media. Part of this fear of engaging with the consumers is about losing profit margins by enthusiasts who judge wine quality in terms of price when there has been no other evaluation method. Wine in glass bottles is no longer sustainable or acceptable to enthusiasts who have an 'eco conscience' and most of them now do.

Wine enthusiasm will save lives, catalyse social communities and could even revolutionise agriculture. The wine industry could take the lead and cover all vineyards in healthy living soil rather than the dead soil and monoculture it currently manages which needs ever more fertilisers and dangerous pesticides. They could in fact initiate a movement that would eventually trap all the excess CO_2 in the atmosphere which is causing global warming using a natural easy to make substance called biochar. These are just some of the exciting stories you will read about in this book.

You know, the wine purchasing power of all us enthusiasts could save the planet by opting for wine with more eco-bling. We could save ourselves in the process due to the substantial health benefits. We, the wine enthusiasts, are the wine world and we can conserve it for future generations but we need to take action now. Our green purchasing power can shape the future of wine. This is why we need to become more involved and informed about the wine we use or should use in our daily life.

As an Irish person, I am also inspired by the effect Irish people have had on the wine industry the world over. The Irish diaspora such as the Lynch Chateaux in Bordeaux, MacMahon in Burgundy, the Celtic Saints who started vineyards in Europe, the Barry and Horgan families in Australia, Francis Mahoney the 'King of Pinot' in the Nappa Valley and Jim Barrett, the owner of the Californian vineyard Chateaux Montelena. Barrett produced the Chardonnay that was the wine that stunned the critics at the historic Judgement of Paris - a landmark blind tasting that changed the world of wine forever. Wines of calibre, after this event, were accepted to be in the new world as much as the old. I hope that this book I have written continues the tradition of the contribution of Irish people to wine culture.

The wine universe is very exciting, interesting and well worth your exploration through the lens of

NutriWine. So here's to your health. *Sláinte!* That's Irish for a toast to your health and wellbeing.

1

The Wine Mind

IF you're a wine enthusiast it's really important that we establish who you are - as YOU are the world of wine.

Would you believe it takes you 38 seconds to decide what bottle of wine you want to buy?

You are particularly attracted to wines with medals on their labels. What the psychology of your mind appreciates about wine in terms of seeing, tasting, smelling, thinking and most of all buying, perhaps even believing, drives the industry both in culture and innovation. Psychology Professor Larry Lockshin found that it takes you this time on average to buy your bottle of wine and much of that decision is based on what the label is and its design.[1] More than ever we want wellbeing, escape from stress, relaxation and meaning.

Wine appreciation delivers on all three. The world of wine is easy to understand, enjoy and belong to with a few basic pieces of information. The first place to start your journey is to know who exactly drinks wine and why? After all, in the USA there are over 80 million of you.[2] Globally over 26 billion litres of wine is consumed annually.[3]

The Wine Enthusiast

Every month you wine enthusiasts drink wine in:

- the USA once a week
- the UK once every month - lots of women drinkers
- Australia 21 glasses per month[4]
- China at the weekends socially by young professionals or at banquets.

You would all drink more, you say, if you knew more about wine and could easily get that information in a way that was not education. If you're a man wine is about status and success; if you're a woman then it's about socialising and in fact 80% of all wine now purchased in the UK is by women.[5] If you have an oenological 'instinctual drift' or wine mind in the U.S. then you will be living generally in the east coast and have a professional career.

The three words that are key to being a wine enthusiast are colour, smell and taste. People who like to explore wine for relaxation are as much enthusiasts as the ones who love the world of wine as a pet subject or who become sommeliers. After all you are the ones buying all the wine. The world of wine has lots more than the history of the Chateaux to offer you. There are wine pairing, health benefits, new wine makers, technology and a green organic movement full of eco-bling. Something for everyone.

You don't have to make tasting notes to be an enthusiast; you just have to be interested in taste. Don't think the world of wine requires you to swirl wine and start spouting poetic descriptions as such as 'dark as Madagascar' and 'flavours unravel like a striptease'. With a bit of exploration, like knowing the wine regions and grapes, you could deepen your choice. Right now the vast majority of you base your purchasing decisions on price scale. High priced wines are better, and cheap wines are just that. But are they? What about the taste? Aroma? What region? What is the grape? What is the *terroir* or land they have been grown in? Have you ever asked yourself what wines you actually like? Taking some time to explore this would be great. As the ad says - 'You're worth it'.

You might like wines from a particular region like Bergerac rather than Bordeaux. Your wine may be cheaper on the scale but that does not mean cheap in 'experience' which is what you truly value if you are a millennial. The flavours will all be there and by a great winemaker and vineyard too perhaps. Price does not reflect totally on quality or individual experience or reflect the meaning of the wine for you or your friends. Price is not a guarantee on quality totally. You have to develop your nose and connect with some great winemakers and vineyards.

By exploring different wines, imagine, you can taste the world. Even if you have a regional wine favourite one can still explore and find other likes from a whole range of wines. Plus you get to socialise and have fun. If you always drink what you always partake of in terms of wine that is all you will get. Go lateral - as the mind once stretched never regains its original shape.

Generation X & Millennial Digital Natives

There is one of two groups you will generally belong to with your wine enthusiast peers – the Baby Boomers and Generation Xers or 'Millennials'. If you were born before 1980 you're in the Xer's 35 – 46 age set and if it's after, you're with the Millennials. The two groups have very interesting psychological outlooks when it comes to wine. If you're generation X, then your appreciation of wine is with food and dining. Wine is the reward at the weekend for all the hard work and helps you to relax and unwind. As you get older you seem to get into wine tasting more and deepen your sensory understanding. You do seem to slip back to beer but some good reasons, starting perhaps with health, could keep you in the vineyard.[6]

Millennials are the darlings of the wine industry. Marketeers everywhere are bouncing up and down about them. You love new, go for adventure and love 'experience' products. As a millennial you will be under thirty and will already have an appreciation for the taste of wine as it will be your first legal drink. You will use wine as a lifestyle choice and in social settings, and as you age and settle down it will become part of meal times, enhancing your dining experience. The industry is getting very excited about you Millennials but seems to be doing little to engage with the 35-54 age bracket who are responsible for 44.1% of all wine purchases. This is double what the Millennials buy and as the Boomers (45-65) age, they are more and more likely to drink wine. As a Boomer you will account for one in four bottles of wine consumed in the USA. The Wine Department of the Silicon Valley Bank recently stated that the Boomers in fact drive the luxury wine market.[7]

All of you wine enthusiasts are also very smart. A third of you have completed graduate school.[8]

Where the Millennial generation is starting out is different from Boomers and Generation Xers. The Baby Boomers were born just after the end of alcohol prohibition and were the children of men and women who grew up in that era and through The

Great Depression which did much to stigmatise wine appreciation and culture. They dined with wine but only started appreciating wine and socialising with it later in life. Much of the wine they drank was actually made in the garage. For them it was something that went with food and was part of traditional family life, especially if they were Italian. There were no sophisticated Chateaux or wine regions. If it came from the garage or cellar there would have been little art and craft in the wine such as we see in wines today like Pinot Noir. There were no wine tasting evenings and wine was not used for socialising but dining. The wine health connection did not come into the everyday vernacular until much later in the 1990's.

The Millennials clearly start wine appreciation at a very young age. They view wine with leaps of value as a social product which satisfies their needs for sophistication and also has value in terms of luxury and health. Health is a major trend you follow. Over 20% of males in this bracket drink wine as opposed to 6% of the older age groups. Half of them, nearly 36 million, have yet to reach the legal age limit to start appreciating wine.[9] Being the first generation of 'digital natives', in other words computer savvy, they appear to break beyond the fear that blocks generation Xers in exploring wine through the internet. They ask their wine questions online and get immediate answers. This digital space allows them to ask the questions a Generation

X person is afraid to ask for fear of embarrassment. Social media is all about conversations and Millennials understand that new questions create new answers that generate new possibilities especially in their choice and experience of wine.

They like to be part of a movement generated by the other enthusiasts rather than slick ads. This is what gets them all twittering. They trust their peers' evaluation more than an expert or critic. In other words they tell you what wine they will be drinking. They also buy the higher priced wines as a luxury wine. You see they are already onto wine sophistication through sensation: colour, smell and especially taste. They want quality not quantity, and experience not bumper value packs. They are curious about wines from other countries like New Zealand, Australia and Chile. Their wine choice does not suffer any xenophobia. However, even though Millennials are transforming the wine industry through social media which influences news feeds, Generation Xers are also being encouraged to explore their wine and expand their range of taste and experience too. The Millennials are actually fuelling the growth in diversity right across the generations. How exciting and even more so when you consider what power they have to shape the future of the wine world.

In reality, the wine and health connection is going through a sort of renaissance from the damage of prohibition and viticulture is now being truly valued as a craft and science beyond the bottle. Wine culture certainly has an allure or x factor which the Millennials are attracted to. Certainly the young professional Chinese. The new wine appreciation movement taking place on social media is rapidly revolutionising the world of wine for people in every demographic and the industry too.

One thing no one seems to have thought about is how to increase the amount of black people drinking wine - a minority who only account for 20% of wine drinkers. In addition over 30 million Millennials are Hispanic and will need to be marketed to in a specific cultural way. In both cases there is a danger that if the industry is not mindful they may lose entire generations to beer.

The Wine Lady - Entertainment & Elegance

If you're a lady living in the UK you purchase 8 out of the 10 bottles of wine in your home.[10] However you only drink wine once a month. If you're a young professional female in Manhattan then you are more than likely drinking foreign wine - Italian in fact - and ordering wine by mail order. Over 80% of the mail orders for wine are by your good selves which may surprise people. Women purchase nearly 60% of all wine in the U.S. You may have 12 bottles on the wine rack on average and will be the one to pay $15 and above for your wine. Women are pushing sales in both sparkling wine and rosé globally. In the last few years the glut of rosé wine can be clearly seen increasing in the supermarket aisle. When confronted with the wine wall in a supermarket perhaps it's just easier to grab rosé and head off to gather the rest of the weekly shop. As women are loyal to their brands both of those sectors of the wine industry are booming. Women want to entertain and sparkling wine helps create a bubbly ambiance with friends for parties.

On the other hand women are fast becoming connoisseurs openly. The wine industry, looking inward as usual, has been marketing to women through the creation of kitsch feminised marketing. Like the pink labels and even the sweetness of the wine. Most sensible women would not touch that stuff with a long barge pole. Women know when they are being given a hard sell as it's they, 75% of them, that do the shopping at the supermarket. There is no meaning other than profit for the producer in these generic wine products. When a women goes to buy wine it's not an impulse decision, it's well thought out and needed to enhance her time with her friends, family or lover. Wine feminists are beginning to be very vocal about this whole 'girlie' marketing issue and clearly feel they are being patronised. Many women are more knowledgeable about wine today.

A study in the UK found that women believe that buying wine with groceries was not really 'buying' wine.[11] They felt that it was the man's job to get the specialist wine and even order it in the restaurant. Meridian wine makers are helping to change that awkwardness with their publication '7 Things Every Gal Needs To Know About Ordering Wine'. Women, wine marketeers believe, buy wine for a moment whereas men tend to hoard the wine. Ladies shop with their experience in mind. For example Italian wine with Italian food. Men are looking for high ratings perhaps a trophy. These are just general observations and in time this will change with the female Millennials.

The keyword for women and wine has got to be elegance.

Chinese Wine Aspirationals

China is booming and fast developing a thirst for premium wines. Wine is part of a trend towards luxury goods there. With an ecosystem of a million millionaires their thirst for fine wine will only increase. The Wing Lung Bank in Hong Kong recently started to offer loans of over $600,000 to build wine portfolios from a select list of over 50 wines from the Bordeaux region. Hong Kong has removed taxes and importation duty on fine wine. Most of the wine you buy is red and the colour red has lots of meaning in your Chinese culture, not least as being very lucky. Fine wine is a status gift for you. Wines are repeatedly reaching record prices at the Hong Kong auction houses. Every year China drinks 1 billion bottles of wine and has a market that is growing by 30% annually.

If you're a new Chinese wine enthusiast, chances are you're one of the new jet set entrepreneurs who are into sports cars, designer clothes – all luxury goods that include wine. You are a wine neophyte but wine gives you prestige. Luxury *putajau* is the badge of your sophistication. First growths of Bordeaux are what you have been focusing on. You drink fine wine as part of your lifestyle rather than hold onto wine to invest. You like to be seen drinking wine at fashionable bars but you don't quite understand wine appreciation. You want any wine as long as it's French which for you is the height of sophistication. You are a sophisticated consumer and you will easily pay for an 'interesting' wine.

You are all young and excited about wine. However you find it hard to get good information. As regards wine tasting, flavours like blackcurrant are culturally foreign to you. So you need reliable sources to explain wine culture to you and perhaps culturally translate it. As you are growing exponentially, this is wine history in the making and is the responsibility of everyone involved that your own wine culture *premier cru* grows in the right direction. Your lack of wine education can lead to you being ripped off. You may end up paying $100 for a $1 wine relabelled as Bordeaux or Burgundy. Counterfeits abound. You want to drink wine as you don't want to be a *baijiu* drinker which is a traditional grain spirit. In fact the Chinese government has been trying to encourage people away from its use and are promoting wine culture as a healthier alternative. Every year in China over 2.5 million people graduate from university most of whom will become wine drinkers.

Simon Tam is seen as the key contact for wine in Asia as Head of Wine China for Christies. For over 20 years he has been educating the Chinese wine market. Hong Kong Futures is becoming a hub for Asia. All of the great names in the world of wine have been attending this new event. As the market potential in China is so huge wine making will become more and more popular in China. Yao Ming, the Chinese basketball player, has launched his own label of Californian wine. He has strategically positioned himself to welcome all the Chinese interest in wines of the Napa valley. What a hook shot. He is clearly a visionary and entrepreneur and will be in position when the Chinese interest in wine will naturally evolve to California after their entry via Bordeaux to the wine universe.

The fine wine makers in France want to put a ceiling on what they export to China. Some Bordeaux wine producers have capped the levels of their stock to protect their wine and culture being entirely exported. This could be some crafty economic oligopoly too. Your market is so huge you could swallow what they produce in one national gulp. Some of you are even beginning to buy your wine by the vineyard.

Wine speculators need to remember that standard wine is imported and mixed with Chinese wine and there is a very low profit margin. So the way to profit from the Chinese market is through the luxury wine route, even though the Chinese wine market is expected to be the biggest in the world in the next few decades. Chinese people also may take easily to wine tasting as they have a sensitive nose. If you look at the world of perfume they like soft scents and therefore can notice and appreciate delicate aromas.

Chinese wine making is about to explode in production and innovation. Some years ago when I attended a conference on stem cells the western scientists told me that when they visited labs in China they were amazed at the dexterity of the Chinese scientists in working with stem cells. Wine will be no different for the Chinese especially if they start moving into luxury wine production. When they figure it out with government support. This century will belong to China.

Future Trends & Innovations

We as wine enthusiasts are shaping the new trends in wine. Man, woman, Baby Boomer, Generation Xers, Millennials, the young Chinese aficionados. All of us. Our minds and our perceptions are influencing this global industry. Over the next decade we will want more information about our wine as we will firstly be more informed. Eco packaging will be something that will impress us, as well as wines with eco-bling. There looks like a beginning of a BYOW (Bring Your Own Wine) movement to restaurants where they will cork it for a charge. More people will be exploring the world of wine through trial size bottles the size of a single glass of wine. Hopefully with this book people with medical conditions will be encouraged to start moderate wine drinking to help extend their lives.

Wineries may be able to sell more wine direct to their consumers. People will be storing more wine in their houses. Wine appreciation will move away from snobbery and critics and towards an open and inclusive culture that is based on experience and moments of meaningful living. Wine is going to be discussed in terms taste, quality and perhaps tradition but even more so what your wine label means to people. The fans of your wine label may even create a new meaning for it. There have been a few cases of this happening on Facebook about which people say that if it was a country, it would be the fourth biggest in the world. More and more of us will be talking about wine and when we get to Wine 3.0 this will have such a huge impact that none of us can predict what the outcome will be. This is the next stage in social media when the www (world wide web) goes semantic.

Chinese wine quality will improve through joint ventures. More people will be buying bag-in-a-box containers as glass bottles are just not sustainable. The world will be awash with sparkling wine - all types - not just Champagne. Wineries will start talking directly with their consumers through social media. Wines with a lower alcohol content will be in increasing demand. Millennials, as social groups, will start making fine wines in the garage creating a demand for technical kits and expert advice. Wine brands will need to have a narrative to succeed. New technologies will have to be invented in viticulture to reduce alcohol levels and protect taste in wines which are being affected by climate change. Organic wines will be more on the consumer purchasing radar. Enthusiasts will go to classes to learn more about wine but they won't be in traditional settings. More wine tour tourism. You're going to be drinking wines from every country as part of your exploration and fine tuning your palate. Most of all wine will be increasingly connected to health and wellbeing. Innovations and new product developments will take place around this trend.

So there is a lot to get excited about and many different doors for someone to enter the fascinating world of wine.

2

The Tasting Lab

Your nose has over 15 million sensors to experience the aromas of wine. The way in which you then cognitively order them in your mind is what makes you an expert.

The whole point of wine appreciation is to come to an understanding about the different characteristics of wine. To have a knowledge of the tastes, colours and aromas of different wine styles. This also has significant benefits, as the longer you hold the wine in your mouth, you absorb 100 times more the anti-ageing phytochemical resveratrol than you will by the stomach. Sommelier Raj Parry gives the best guidance when he says, "Taste what's in the glass, not what's in your mind."[1]

See Swirl Smell Taste & Savour

When you come to wine tasting you just need to remember five S's which are: see, swirl, smell, taste & savour. See the wine's colour, swirl the wine to release the aromas, smell the wine but seven times, taste but swash it around in your mouth and swallow and savour any aftertaste.

Wine appreciation is normally known as wine tasting, which should have really been called smelling, as this is the primary sense you'll be using. Wine tasting is similar to learning a new language and all the benefits that can bring in expanding your mind and life experience. You start off with the basics - the wine varietals - Cabernet, Chardonnay or Pinot, and build a vocabulary around those of 12 categories of scents, two examples being fruity and woody. When people go and learn a new language, one of the main motivators is that they have fallen in love with someone. Many people fall in love with wine; once they touch the vino mandala they fall deeper into wine culture.

Learning Chinese is more than just the language, it's about the people, geography, culture, history, innovations and new discoveries. So it is in the wine universe. This is more than just a table wine experience there are so many flavours and enticing odours to encounter. This is also a way to connect back to the earth through the soil in which the vines are grown. By learning about the regions and having a different wine type every week within a time frame of six months you will broaden your tasting experience and increase your choices and options. Not to mention probably also get a lot more quality for your dollar.

Wine appreciation does not mean by any means that you become a wine bore or snob. By getting a handle on the art of wine, how it's made and what the different features of wine are, you can have an interesting journey that will benefit your brain, diet, life and health. You get more out of it. Most importantly, there may be a varietal that is much better than the one you always choose. The truth is that even though there are lots of wine labels, wine falls into a number of categories that we can easily handle, more so today with the availability of apps on your smart-phone. A few tips first.

When a waiter comes to your table and pours some of the wine you ordered into your glass, don't sip it. This wine etiquette comes from seeing if your wine has corked or not which means spoiled. Rather, swirl the wine and then smell it and as long as the wine does not smell like soggy paper then gesture to the waiter to continue. Tasting the wine shows you don't know that step. When you enjoy wine this does not mean impressing your date by doing it the professional way at a conference - slurping and making vacuuming sounds. This is only needed in wine education not at a restaurant unless you're Steve Martin in a comedy. A wine aficionado may do this very very quietly otherwise the dining area could turn into a Monty Python film set.

The first step in wine tasting is to swirl the wine in the glass. This allows oxygen to get into the liquid and to break open the flavours which will evaporate with the alcohol and which you will use your nose to sense. The more the oxygen gets into the wine, commonly known as being allowed to breathe, the more the aromas will unwrap themselves and their alchemy.

The Wine Rainbow

One of the features of wine appreciation that you are first taught about is colour. Wine ranges in colour from golden yellow right through to ruby red. Wine colour can tell you the age of the wine - even the freshness - and can be a reflection of the wine's character. White wines gain colour as they age whilst reds do the exact opposite and loose colour through oxidation. Young whites will have a slightly green hue and five year old wines a yellow colour with a thick 'legs' when you swirl it around the glass. These legs are telling you of the sweetness from the higher levels of glycerol. White, left to age, becomes what is known as 'maderised' which looks like a brown ruby colour. Red wine on the other hand is generally made from two types of grape - thin and thick skinned. When red is a young wine you will see the red colour in the 'legs'; as it matures it does so towards a tawny port colour. See? It's all very easy. Now when you walk the aisles of white wines at least you can know the ages of the wines based just on colour. Scientists now know that the compounds responsible for much of red wines' colour are anthocyanins. They have found that the way they bond with other wine chemicals can even create a blue affect.

With novice tasters it's been seen that they are influenced by the colour in terms of taste. When researcher Dr. Jeannine Delwiche was Professor of the Sensory Science Group at Ohio State University before moving to Princeton, her team coloured white wines to look like rosé and found the novice tasters were very influenced by the colour.[2] Wine experts were not taken in by the colour doctoring in another study in New Zealand done by Dr Wendy Parr. When, despite the wines having been coloured, the experts relied on their tasting experience and training. They were right - unlike the novices. So, even though colour is an important point, taste experience is more reliable according to scientific research.[3]

Taste & the Electronic Tongue

When it comes to the tasting of wine with your tongue you need to both taste the wine's sweetness or acidity and the sensation the wine creates on the tongue. The tongue not only tastes but feels the textures of foods and liquids. Texture, by the way, is due to the fat in food according to food technologists. Great wines have great mouth-feel, and taste will be as a result of acids in the wine whereas sensation will be as a result of the tannins. You can discover the body of wine - basically its subtle viscosity - in how the wine feels in terms of weight - light or heavy in the mouth. When the wine flows across your tongue the texture gives a silky, velvety or buttery sensation. Sometimes even a chalky feel if the wine has minerality. The tongue also allows you to experience what tasters describe as the wine's 'finish' on your palate. A great finish is when the tastes and the flavours last on the tongue after you swallow the wine.

Your tongue has roughly 10,000 taste buds that allow you to taste generally five tastes: sweet, sour, salty, bitter & umami. In five areas of the tongue they are separately depicted and have the same surface area. You also have taste buds in the soft palette and some in the throat. Some scientists believe that all the taste buds fire neuro-electrical signals that recognise all tastes. The tongue receptors also have a taste threshold that most wine supersedes which is why there is more reliance on smell in the practice of wine tasting. The individual production of saliva also either increases or decreases taste sensation.

Saliva is slightly alkaline and therefore cleans the tongue and wipes the tongue clean of taste stimulus. This allows the tongue to calibrate to a neutral tasting threshold. People who are stressed have less saliva in their mouths which is why it's always dry. So noticing your saliva levels will tell you a lot about your stress levels. Normal saliva levels will help you taste better. If you watch closely how Chinese people dine you will see that they eat rice between mouthfuls of ginger beef or stir fried chilli chicken. The reason is that rice alkalises the tongue which gives the Chinese a superior dining experience in terms of taste. The rice pH resets the tongue to zero after eating a cooked dish. Each time they taste the main dish it's as if it's the first time again and again. Try it?

Tongue wine experience is a dance between the acids in wine, neurotransmitters and the alkaline saliva. On that basis wine really adds to a dining experience in the level of taste and flavours that will be fired back to the brain electrochemically via these neurotransmitters. In a world of sweet-tasting sugar overload and sour-tasting junk food, you can see why wine tasting could improve a person's quality of life by helping them sense more. This would encourage them in terms of food to eat healthier choices.

Tongue feedback on wine will tell you whether a white is sweet, dry or even buttery. If it's bad white wine, it will taste like boiled vegetables or worse. The tongue will also tell you most importantly about wine's acidity which is the 'nerve' of wine. Red wine on the tongue will have more texture as well as flavour. The texture will range from soft and velvety to hard and leather-tasting depending on the tannin levels. Tannins are also found in tea and give that dry earthy effect on the tongue depending on how strong the wine is. The taste of sweetness from berries in reds and the citrus of whites depends on the tongue and the alkaline saliva balance across 10,000 taste buds exploding like fireworks.

Wine connoisseurs understand the balance in wine through the acidity. The main acids in wine are tartaric and malic. The acid in wine gives the wine its *élan vital* - too much and it's sour, too little and it's flat. The acid triggers taste buds which in turn trigger the production of saliva which will help you chew and digest your food. So between each sip of wine, as you dine, both the saliva and food reboot the sensory slate for the next wine sensation. The relationship between saliva production and wine acidity is rarely referred to in wine tasting. This dynamic could be causing an expanded gastronomic experience in terms of taste. Another great reason to have wine with your food to expand the quality of your dining experience.

The acidity on the tongue could also have the same effect as bitter herbs on the digestion. These herbs are known to help stimulate the production of digestive juices including insulin, encourage appetite in people convalescing and act as a digestive which also protects the stomach lining. So you can see why hospitals should also serve wine. Many already do. Wine must also stimulate the vagus nerve which is connected to the stomach and sensitive to stress. Scientists now believe that a person's wellbeing depends on the vagus nerve's vitality. When it's switched on so is the relaxation response which will reduce your blood pressure and heart rate and keep you out of fight-or-flight biochemistry which can be damaging. According to Tibetan Medicine - the oldest medical system in the world - stress is a form of self-attacking.

The acidity in wine plus the gastric juices it stimulates, particularly in the stomach, can aid women with osteoporosis. The stomachs of over 44% women turn alkaline during menopause and stay that way in post menopause. In order to absorb calcium, the body needs it in a form called calcium acetate not calcium carbonate which is the supplement form. Acidity in the stomach provides this and this is one theory as to why osteoporosis develops in the absence of acidity in post menopausal women. Enjoying wine can reduce bone deterioration for this reason.

Researchers in Barcelona have a special penchant for electronic wine tasting tongues. The university has developed a computerised tongue that detects different types of cava or sparkling wine based on sugar levels and mathematical algorithms. Google was developed with an algorithm. Cecilia Jonquera-Jiménez and her team at the Institute for MicroElectronics (IME) developed a tongue that is able to taste wine and tells its grape and vintage. The device uses synthetic membranes connected to a silicon chip and each membrane detects a specific chemical. This electronic tongue can be further upgraded to detect a whole range of chemicals in wine which is a complex subject, as you can imagine. This could help in terms of quality control and standards of wine making whereby the tongue would screen wines before they got to tasters to ensure projection bias was not taking place or ever could. This robotic tongue from IME measures acidity, pH, alcohol, sugar, potassium and ions.[4]

These new technologies will be able to do what all humans no matter how trained and experienced can do. This is the potential. However the robots will never make wine an art the way humans do in a passionate way. These machines are needed as quality control devices. We will essentially need both in the future and these devices may be one of the future tools in how we educate people about wine. This could be quite fun to taste and to see how correct you are against the computer. A device for what wine makes you feel is just not possible as it's not a mathematical equation it's a way of living and individually feeling.

Smell - Vino Aromatherapy

When you have swirled your wine and the aromas have started to unfold, you need to take seven inhalations of the vapour arising in order to truly smell the aromas wine possesses. Slowly that is. Even when you drink wine the exhalation blows wine aromas from the mouth to the nose and the olfactory receptor. So it's important to savour and breathe through your nose. You know, it does not matter if you can say what you smell at the start, it's more important that you notice what aromas you can connect to, build on those. Some scientists would say it's better not to label as you will end up in an area called 'verbal overshadowing'. This is a nice way of describing someone who is just intoxicated by the fluency of their verbosity. For them they are only experiencing their projection of what the wine is not the reality.

There are three aroma types in wine: primary, secondary and tertiary. Primary aroma is referring to the aroma created from the grape varietal. Secondary refers to the bouquet created from the fermentation and the oak ageing. Tertiary aromas derive from the ageing of the wine in the bottle.

Natural fragrances have a major effect on the receptors in the limbic part of the brain and therefore can improve emotional mood. The brain has a profound memory for aromas. The alchemy of wine is in this zone for smell. There are over 700 aroma chemicals in wine. Smell is one of the most interesting parts of wine appreciation. The nose olfactory nerve is connected to the limbic part of the brain which happens to be the emotional part of the brain. This is nowhere more appreciated than firstly in aromatherapy and perfumery. To be able to taste wine fully you have to be able to go beyond taste which is limited in range by tongue receptors. These

700 chemical wine compounds are called phenols and these also affect the taste, colour and mouth feel of wine. If you want to get more out of wine start appreciating the aroma.

Synthetic fragrances are made from amines and these amines do not have the same effect as the aromatic quality of wine. Synthetic vanilla aroma is one chemical amine whereas vanilla essential oil can have up to 30 different aroma compounds. People who live in cities and who love wine can also appreciate the smell of the wine region through the bouquet. In heavily populated areas polluted with fumes this could do a lot of good where wine not only addresses Dionysian starvation but also sensory deprivation in terms of smell which will depress the mood. The space agency NASA is well aware of what sensory deprivation does to humans. So much so they send astronauts to space with vials of essential oils one of which is basil. There are no smells in space.

Margarete Maury, the founder of aromatherapy, said that essential oils are the purest form of living energy that we can insert into man. The chemicals that wine aromas are made of are based on the same structure some plant aromatic essential oils have. Her husband was also famous in health as he was a doctor who prescribed wines and wrote a book to promote this in medicine. He also knew the power of natural plant compounds in promoting health and wellbeing. These extracts are the core of the plant's being. To get an idea of how much so, remember it takes three tons of rose petals to make one litre extract of rose essential oil. These natural compounds are also in wine in the form of aromatic compounds.

Detection Theory & Wine Judges

Wine smell is the current area of research into wine sensory science. Researchers have started with the nose and in particular the olfactory receptor. Wine sensory research started out in the 1950s but there has been little research since in comparison to food products. The faculty I studied at based at my own *alma mater*, the world famous University of Reading UK had its own sensory research lab for foods and drinks for the industry and sensory research was constantly taking place. This area of research is very important when you consider what a 90 point judgment from Robert Parker does to a wine and all the mini Parkers out there. Our ability to detect a smell has done well in terms of areas that have been researched but our ability to asses cognitively has been practically ignored.

In the last decade or so scientists are investigating why even experts can only distinguish between four separate odours from a complex bouquet. However when they are given more than four individually they are able to name them all. In terms of wine smell we all prefer to identify a mixture, like face recognition, rather than separate individual components. The best wine experts have been shown to have an edge on perceptual skill in terms of how they order their thoughts and recall taste memories. However we all have the same level of smell ability and potential.

Asnomia is the term to describe a person's inability to smell or even taste certain aromas. Something like a blind spot in smelling and tasting ability of the body. We all have smelling blind spots to some degree or other which is interesting to note and makes the individual tasting even more interesting to see if you or your friends can smell the same aromas. If they can smell cherry, apricot, peach and apple and you don't smell the cherry you may have cherry asnomia.

Detection theory has been used in a number of different disciplines and is now being investigated in the area of wine tasting.[5] Detection theory in psychology is the study of the decisions people make in situations of uncertainty such as studying what pilots would do in turbulent weather. In those circumstances, one can measure what is called 'response bias' and this is also applicable to wine tasting. Could even an expert transfer his experience of wines onto others? Those with an untrained nose will probably copy language they have heard an expert use but there is no cognition in the frontal cortex of the brain. They will be using the emotional parts of the brain.

Clinical psychology is for two types of people - those who are too highly strung and need to tune down and others who need to be tuned up. Two very important parts of how we see the world or how they are clouded by our ego perceptions are called projection bias and transference. Knowing how these two work will even transform some peoples' relationships. Projection bias is when we project onto people our own unwanted feelings and issues. Transference is described in psychoanalysis as when we redirect feelings of one person onto another. This also has relevance in wine appreciation and judging and needs to be studied especially the whole area of bias. People can project onto wines irrelevant information as much as they do onto people without basic wine education and tasting experience. This is why China needs wine education courses as soon as possible or they will start trying to project onto all wines attributes exclusive to Bordeaux wine.

Every person according to NLP (Neuro Linguistic Programming) is more capable in one of the five senses. As your sense of smell is connected to the brain there is a case to be made that the more you start to smell wine and appreciate and learn the range of tastes you could possibly become more emotionally intelligent as the olfactory nerve is connected to the limbic part of the brain which is concerned with emotion. If you start using these parts of the brain more, expand the range of scents, you could also evolve emotionally even perhaps balance the brain itself and contribute to your wellbeing. But is brain activity mind activity?

There is lots of psychological research to show that smell has an effect on mental processes and our behaviour. Why else would British Airways have used essential oils in their arrivals lounges? Why does the Conrad in Bali bask in a cloud of lemongrass as a little perk for jetlag? Science also knows we have difficulty labelling a smell. We tend to have a memory trigger rather than the right name. Basil essential oil will remind people of baby gripe water, which their soothers were dipped in to get them off to sleep as infants. Our memory connections with smell can go deep into the brain.

Wine Aroma Wheel

The Wine Aroma wheel was invented to assist people in terms of understanding and remembering wine aromas. The wheel, you could also say, acts as a memoronics tool to help people get started on the range of smells there are in wine appreciation. UCDavis Professor Ann Noble saw the problem that whilst we have words to describe primary colours we don't have this for smells. We had no aroma visual cues. When we smelled orange aroma in wine we had no clue and had to recognise this without the cue. Being a wine scientist she knew what the exact range of wine aromas were and from this devised the Wine Aroma Wheel to help people have cues for smells. The wheel is a catalyst to develop the sense of smell for wine appreciation. The wheel is also rainbow colour and the smells are categorised into three sets I will call primary, secondary and tertiary. Practice with the wheel will enhance the cognitive ability to discern between different wine smells.

Starting within the centre circle there are 12 primary smell classes to place the wine into, herbaceous, fruity and woody to name a few. If it's Sauvignon Blanc for example, it will immediately upon smelling, fall into herbaceous. The second circle of terms within the herbaceous spectrum to decide from concerns whether the wine is dried, canned or fresh. If it's Sauvignon more than likely it will be fresh. Then you need to decide from the third circle what aroma it is. Cut green grass, bell pepper, eucalyptus or mint? So it's a three step process that will enhance not so much your range of smell but the ability to categorise smell correctly, with what could be described as the Wine Aroma mind map. Mind maps were made very popular by Tony Buzan who views them as a vastly superior form of memorising data. In fact he believes mind maps work in a way that uses both of the brain's hemispheres. When you consider that the sense of smell itself is located in a different part of the brain from that which cognates and categorises that sense of smell, you can see why the Wine Aroma Wheel has been so popular, necessary and successful. A stroke of genius.

The Robot's Tasting Notes

An interesting robotic development is called WineBot from NEC technologies in Japan.[6] This little robot also has a penchant for wine tasting using an infrared spectrometer in one of its arms that gives different absorbency readings for different wines. Each wine will have a different chemical fingerprint. The robot can currently distinguish only a couple of dozen wines but in the future it could be used in quality control; and for the rarer wines it could be used to test for authenticity without opening them. The idea is that eventually you can talk to the robot - this one already talks - and it will choose the wine for you. At one point it described a journalist as a sausage. Maybe in the future we will talk to robots about our likes and dislikes in wine and the robot can choose a wine from the cellar based on that information. Wine enthusiasts and experts believe that it can never overtake human tasting nor the enjoyment that gives. However it may also help in the sensory research to see how much 'verbal overshadowing' is taking place in wine tasting experts. In the future we may see the experts arguing with the robots especially as more artificial and cybernetic intelligence develops.

The Sommelier's Brain

Just to prove to you that wine tasting is not made up and that wine professionals can have brains as well-tuned as the best perfumers or expert antique dealers, let's take a look at a recent study that was done on them by Castriota-Scanderbeg at the Santa Lucia Foundation in Rome.[7] Researchers studied seven sommeliers and seven 'controls' using Functional Magnetic Resonance Imaging (fMRI) during wine tasting. With the sommeliers, they found that a cerebral network involving parts of the brain connected with food, smells and taste called the left insula and orbito-frontal cortex were activated. This means that the sommeliers were analysing wines from their database of wine experiences they previously had from training. In stark contrast to this, the controls – who were neophytes - were using the emotional parts of the brain. Therefore the control group only projected ideas onto the wine they were tasting, as they had no database to dip into, through lack of training.

In addition, the sommeliers activated the pre-frontal cortex which is taken up with thinking but the neophytes failed to do this and instead activated parts of the brain connected to emotion like the amygdala. What is also interesting to note is that the neurons connected to the amygdala are very rudimentary, in comparison to the tight construction of the networks in other parts of the brain. The amygdala gland lies just above and in between your eyes and releases dopamine when a person is stressed which causes the frontal cortex to jam you could say. This is what happens when you go blank at an exam. Or when you are under severe stress and just can't think straight. Oxytocin - the love hormone - interestingly reverses this effect giving credence to the benefits of practicing compassion.

The images the researcher saw and the difference between them scientifically, proves that indeed sommeliers are well trained and use a wine educated part of the brain when choosing wines for us and even when pairing wines for our meals. This would be correct as the sommelier has to be trained to know all the tastes and smells of wines, the varietals, regions and labels. So a lot of training and expert thinking go into being a bone fide sommelier. The neural networks that the imaging studies revealed have developed for the qualified and practicing sommelier and are the source of their data bank. A part of a good nose as these studies demonstrate is an ordered mind.

Researchers in the past also did studies on London taxi drivers and found the hippocampus parts of their brains to be larger than usual. This was believed to be due to having to memorise so many streets and routes.[8] Parts of wine experts' brains could have the same extra capacity, like extra memory in computing.

The $20 Million Wine Nose

For an employer to insure any part of your body is flattering. When Angela Mount was working for supermarket chain Somerfield such was her positive impact on their business they called their insurers. Angela had to have a meeting where she was told her nose was worth $20 million to her employer. Achoo. Bless you. The story of the record breaking valuation hit the headlines worldwide. Angela is renowned as a wine expert buyer who changed the wine world in such a way in the UK that high quality wines became available on the high street. Somerfield estimated she was in charge of 10% of their total business such was the power of her sniff. She has continued her wine consultancy elsewhere and has also moved into coffee and wants to improve the taste and aroma of coffee with her record breaking olfactory bulb. How much could her tongue be worth? Marlene Dietrich insured her voice for $1 million and Egon Ronay of the dining guide insured his palate for £200,000. Formula One driving star Fernando Alonso had his thumbs covered for £9 million. Since then other noses have come up for protection but none have been valued at $20 million. That olfactory legend still belongs to Angela.

Wine Pairing & the Molecular Recommendation

The whole experience of wine tasting gets better when you start to pair wine with food. This will also further expand your frontal cortex for wine information and cause more health peaks in the body biochemistry as it reacts to the flavour and the food is even better digested. Generally speaking people pair white wine with white meat like chicken and fish and red wine with red meat, typically beef and lamb. Very sweet wine goes with dessert as any other wine paired with a sweet dish will taste acidic. The whites will cut through brine and the reds will add texture to the meat and both will draw out flavours. For restaurants it may be helpful to consider the components in terms of taste of the wine, as that can't be altered whereas the food can be designed around the wines offered. The elements in wine are set long before you dine.

The menu could offer a choice between three wines that will go well with a particular dish stating the tasting notes. This may also help those who may find it intimidating to get into a conversation with a sommelier who is qualified in wine appreciation of a wide range of wines. BYOW (Bring Your Own Wine) will also be a good way to encourage wine with food and fairer in price too. This would also encourage people to find out more about wine.

Whites range in terms of dryness to sweetness and age as we have seen earlier. With reds you're talking about tannin and oak from a taste range from soft to a hard taste that extends to leathery. A dry Sauvignon Blanc will go well with grilled fish and crab with its citrus notes. With roast chicken you could have a Chardonnay to complement the butteryness or a velvety Merlot to draw out the texture of the meat. Coming to beef you could have Pinot Noir or a Cabernet Sauvignon - their full body and also tannins complement the meat. Champagne will cut through any brine in fish dishes like caviar and oysters. There are a number of easily printable free charts online to download and print to assist your wine pairing. With the huge popularity of Asian food now around the world think Rieslings from Germany for refreshing crispness.

You might think that all this is pie in the sky but there is a lot of science behind wine pairing. Francois Chartier wrote his bestselling book 'Taste Buds and Molecules' all about the science of wine pairing and how the chemistry of wine goes with food. Francois spent two decades as a sommelier in Quebec but was not satisfied with the knowledge of wine and wanted to explore further. He wanted to get into the chemistry of wine pairing, about why the foods went with certain wines, considering the food and wine chemistry together and what were the best synergies based on similar chemical aromatic compounds found in both. He is unique in that he is a sommelier at a molecular level.

As a man who has reversed his multiple sclerosis with diet he has a powerful voice and knowledge when it comes to molecular gastronomy. In fact the multiple sclerosis totally disappeared as a result of his self-treatment which started his journey into food and wine chemistry. Wine and food, Francois has seen, share exact aromatic compounds. For example, bell peppers contain 2-methoxy-3-isobutylpyrazine which is found in Sauvignon Blanc. So those two would go well together. This is a very simple example of how sophisticated this can get but basically he is matching the food and wines based on their aromatic and flavour compounds. If they are similar they won't clash and instead create a gastronomic synergy.

Lamb, he advises to have with Languedoc wines such as Corbieres and St Chinian. The reason being that both have the aromatic terpine compound thymol which is found in the herb thyme. He serves pork with oaked wine as both contain lactones found in apricots which are sometimes served with pork. He also went on to discover what he describes as 'bridge ingredients' that help foods match with wine in that they reduce the synergistic distance molecularly between wine and food. So wine that goes well with mint will go well with dishes made with mint.

The main goal of wine pairing with food is synergy that increases your dining experience. Like any good marriage the wine must not overtake the food and the food must not overtake the wine. Together they bring out the best of one another. White wine for example will cut through a fishy taste and the citrus notes will complement the taste of the dish. Lemon is generally served with fish and lemongrass in Asia.

There are two paths when it comes to wine pairing. One is by following the advice of experts in terms of what wine to pair with your food and the other is empirical - personal trial and error. Each person will have different likes and dislikes. As Vaynerchuck says often in 'Wine Library TV', if you try a different wine with your food every day, in six months you will know your likes and dislikes. The other advantage is that you will increase your range of personal choice. Wine high in alcohol - as in over 13.5% alcohol - is referred to as "hot" and normally only drunk with spicy food to draw out the spicy hot taste. However many wine aficionados pass on these high alcohol wines as they have a 'hot taste' which means there are very few of the flavours.

Romance, Sex & Aphrodesia

When Pierre-Emmanuel Taittinger addressed journalists at the Reuters Global Luxury Summit and said that the biggest competitor to his Champagne was Viagra - he wasn't joking.[9] Wine tops the list of aphrodisiac beverages. In fact in China herbal 'spring wine' was a major health tonic for its Viagra imparting properties for centuries. Red wine and spirits were mixed with herbs to create infusions with aphrodisiac qualities based on principles of Chinese medicine. The herbs most used were cinnamon, ginger, nutmeg and rosemary. Wine increases testosterone levels in women and therefore libido. The Santa Maria Hospital in Florence studied 789 women and found that women who drank two glasses of wine a day had an increased sex life. All the women had to complete a Female Sexual Function Index questionnaire. The study was reported in the Journal of Sexual Medicine and the researchers figure that the red wine increased blood flow to the sexual regions.[10] More studies need to be conducted, obviously.

Deviation, a dessert wine from Quady Winery, is made with a powerful aphrodisiac herb called damiana. There must be something in the wine as it's been gathering a lot of awards since it recent launch. South American shamans have a long tradition of using the herb as a remedy for impotence. More wines in the future may like to tap into the demand for these wines.

What would be even more amazing is if there was a connection found between oxytocin - the love hormone - and wine. A nine-sided amino acid of love, oxytocin is the hormone that is released by the brain at orgasm and is also connected to post-coital bliss and bonding. People with more of it in the brain also have better relationships. The love hormone has also been shown to reduce stress. Aromas have long been known to induce arousal. Red wines also have hints of masculine pheromones. Apart from allowing you to relax and reduce your sexual inhibitions there are other aspects to the aphrodisiac nature of wine. If you think the power of wine to induce arousal is interesting you are going to be even more amazed at the power wine has in promoting health and wellbeing.

3

Soil to the Glass

$$C_6H_{12}O_6 \longrightarrow 2C_2H_5OH + 2CO_2$$

This fermentation equation is one of the most important biochemical equations. One mole unit of glucose is converted to two moles of both ethanol and carbon dioxide. Through this path all wine and other drinks with alcohol are made. Over 268.7 hectolitres of grape juice undergoes this biotechnology process to become wine each year.

The first wine that came into the world is thought to have been made around 6000 B.C. Since that time wine has been a huge part of agriculture and human life. For many centuries wine was drunk as a safer option to water which was much contaminated. In addition wine was ingested for its health benefits. When you drink a glass of wine try and remember all the effort that has gone into producing it by the vine and that the wine in your glass should be the mirror image in terms of colour, taste and flavour of the terroir in which it was grown.

A wine of great character will have been grown in a great environment.

The Wisdom of Terroir

All good wine makers will tell you that the key to a great wine is the soil, or what the French call *terroir*. There is no direct translation of this word in English but it is known throughout the wine universe. Terroir means the soil type, climate, environment, water levels, even the spirit of the place wine grapes are grown in and that comes through in the wine. Marrying terroir with the right grape variety is the sixth sense that belongs to those who make the greatest wines.

We live in a sapphire blue biosphere of what we call air that covers the entire planet. This is how we are seen from space. We live in an atmosphere of 78% nitrogen, 20% oxygen and 1% carbon dioxide and a few other elements. This ubiquitous biosphere keeps the warmth in and the radiation out. This is what some wish they could see on other planets as it would mean life existed there. The biosphere is only as high as a balloon ride and not as high as people damaging this believe. Nor is it as infinite as some might like to think it is. At a height of just a hundred kilometres up in the sky exists what is called the Karman line - a boundary between our biosphere and the rest of space. The same small distance between Oxford and London in the UK. So you can see why we need to care for our planet now and part of its vulnerability and beauty – the biosphere.

Plant life is a major contributory factor to this biosphere ecosystem and vines are included in the system that maintains this equilibrium: photo-synthesis, the nitrogen cycle, the carbon cycle. From photosynthesis, plants use carbon dioxide and water to produce oxygen and sugar. The bacteria in the soil, if it's living soil, provide nitrates to the plant along with minerals as food. Plant life starts in the soil and the soil has to be living. When herbicides, pesticides and insecticides are used on the land and crops they kill the natural microorganisms in the soil.

Every year in Californian wine country in excess of 22 million pounds of more than 50 pesticide types are applied to vines.[1] Over 17 of them are known to generate cancer. Grape crops are sprayed with pesticides more than any other crop and in far higher quantities. The wine industry perhaps can afford to spray more but should they? The soil becomes dead and you need to pour on fertilisers to feed the plants which would ordinarily get their nutrition via the roots in living soil. When you consider that the roots of the vine in biodynamic vineyards have been shown to grow 20 meters deep and conventional vines by comparison only grow one meter, you can see the damage quite graphically not just to vines but also to nature.[2] Vines sprayed with chemicals have roots that grow sideways looking for food in a dead soil which is not really there.

Apart from the dangers pesticides pose to humans there is also a serious threat to bees. Bees are the engines of nature and cause cross-pollination which generates all the fruit in the world. When the last bee dies humanity has three years of food left. In recent years a phenomenon called 'colony collapse' has been occurring where whole hives of bees die. Beekeepers attribute this to the use of pesticides and in particular *imidacloprid* which has been banned by the French government. In the documentary 'Queen of the Sun' which is all about this bee crisis, one beekeeper is shown moving all his bees to a different state by lorry for fear that they will collapse and die from the effects of pesticides.

This is a serious problem for agriculture and our planet and a good reason to reduce the risk by stopping the use of dangerous pesticides. If biodynamic and organic agriculture do not need to use pesticides and fertilisers, there is a case for its reduction and potential elimination. This will upset chemical company shareholders of course but should we allow them to kill such a major part of nature for profit? As one of the Native American elders once said only when the last tree has fallen, the last fish has been eaten will man realise that he cannot eat money.

One very easy way of getting the soils to be more fertile is the addition of biochar which is what makes the soils of the Amazon so rich and fertile and has been shown to increase crop yields by over 30%.[3] Biodiversity would also help and can contribute to wineries reducing their carbon footprint and in addition reduce the susceptibility to infection. Through the addition of different plants between the vines you increase the type and population of flowers for bees. If you talk with organic wine growers they will tell you that you can control pests this way. Vines grown in living soil and as naturally as possible must contribute to flavour in wine.

The Vineyard Vintage

The vines have to go through the full four seasons - autumn, winter, spring and summer to produce a harvest that creates wine. The year will then be referred to as the wine's vintage.

The care of the vines is a practice called viticulture. Today this 'precision' care is through the use of satellite imaging and other technologies on a meter squared basis. Pruning the vine is one of the most essential parts of this, as by correct pruning you get plant vigour and fruit yield. Vine vigour means a combination of rootstock, variety, climate and soil. The vineyard owner needs to decide how many buds are to be left on for the following year's growth. They then must decide which of the canes will be allowed to grow and this is attached to a wire for support and the others are removed. The remaining cane grows horizontally. Irrigation is now a very hot topic in wineries in terms of the green movement and water conservation. New technologies are helping wineries conserve water more without threatening their vines or crop. In addition if the soil is dead then the soil cannot retain or store the water for the roots. Most vineyards are implementing customised irrigation systems.

As the spring comes and the plant grows the canopy of leaves must be in the sunlight in order for photosynthesis of the plant to optimise fruit production. The leaf colour is monitored and if there is a threat of mildew, copper sulphate or Bordeaux mixture is sprayed on the plant. Vines are managed now in such a way as to generate grapes that have a collective ripeness. In other words, when it comes to the point of harvest all the grapes are ripe and ready to make wine with. To achieve this with different vines and different watering conditions and individual leaf surfaces is the job of the vineyard manager. They have to understand the land and specifically the terroir.

They are constantly analysing acidity, sugar, anthocyanin and polyphenol levels for an optimum yield. At harvest the grapes are picked generally in 20 kg cluster boxes and these are put in 250kg containers and are transported to wineries. This reduces damage to the grapes. If you're going to really get into wine you should plant one vine in your garden to connect with vine. Some people even talk to their plants. Plants have been shown to respond to emotions. You see this sentience especially with plants that trap flies like the venus flytrap.

The Secret Life ofVines

Vines like all plants are truly amazing. They take water plus carbon dioxide and with light and air make sugar. Over 90% of the plant is formed from just light and air. Imagine. This includes all the fruit in world. The blood of the plant is its sap made up of water, nutrients and plant hormones. Some of these plant hormones are quite close to human ones - a good example is abscisic acid (ABA) which is the same as a human cytokine. Cytokines are involved with the immune system. Extracts of clary sage can induce childbirth naturally by stimulating the glands in the brain. These plant hormones inform the plant which way to grow and are called tropisms. Roots go into the soil to find water and nutrients and the plant grows towards the sun looking for light and air. The sap is very important to the vine and lets it breathe and feel the terroir and in order for a vine to grow in a healthy way it needs great sap. You get living sap from living soil. Grape vines used to have a lifespan of 100 years but that has reduced to 30 since the 1960's. Guess what chemicals we started using industrially at that time?

Our world and existence is based on a mutually beneficial system of plants making oxygen that we breathe and transforming the carbon dioxide that we exhale and produce with our industrial endeavours. The protein called chlorophyll found in plants and responsible for its green colour looks exactly like the blood protein haemoglobin in terms of its structure, apart from just one thing - the mineral. In haemoglobin it's iron but in chlorophyll it's magnesium. Magnesium is nature's tranquiliser for humans and over 90% of people are deficient in it. Sea salt is a good source if you need it. If you were ever in a high school chemistry class you will know magnesium is the element that produces the bright light when burned - as bright as a welder's light.

The plant reaches into the soil via its roots for food. If the soil is alive with a food web then the nitrogen fixing bacteria can assist in providing the nutrients. Each handful of living soil has over 1 trillion bacteria in it.[4] The top soil is also rich in minerals but over-farming is destroying this soil and is the reason why conventional produce has low mineral levels. Plants are great sources of minerals such as selenium, chromium and zinc that respectively protect you from cancer, diabetes and immune problems. Plants phosphorylate the minerals in a way that the human biochemistry can absorb much better than supplements. If these are present in the plants, we are healthier for it. Most diseases can be traced back to mineral deficiencies so all the better if they can find their way into wine in natural form. This is why organic vegetables grown in living soil have high levels of minerals in them.

Plants need fertilisers now due to the fact that the soil bacteria has been destroyed by farming chemicals and the mineral rich top soil has been eroded. Climate change is causing big problems for vineyards in terms of erosion especially in Portugal. Fertilisers are made up of macro and micro nutrients such as nitrogen and zinc. Nitrogen is an obstinate element and needs to be fixed artificially by being 'cracked' in huge chemical plants through the Haber-Bosh process. This process fixes nitrogen and creates ammonia fertiliser. The fertiliser from these chemical facilities finds its way into our food. Scientists now estimate that 25% of the nitrogen in our bodies is from one of these chemical plants. The fertiliser from these chemical producers now supports the food production of a third of the world's population. Many water ways have a carpet of algae on them caused by these fertilisers leaking into the water. The algae remove the available oxygen from the water and kill life. As a result of this locust swarms are seen in increasing numbers as there is no fish in the river to eat the larvae due to the ecosystem being interfered with. We now need to be able to feed 6.6 billion people and 60% of the world food production is dependent on fertilisers.[5]

When the plant food reaches the leaves via water that it absorbs from its roots the water is fed into a biological process called photosynthesis. No I am not slurring. Sunlight strikes the leaf and a photon excites the protein called chlorophyll that releases an electron. Two electrons and two protons come together and the loss of the two electrons causes two water molecules to split creating oxygen and the electrons create an energy storage molecule called ATP and NADPH. This energy is used in what is called 'The Calvin Cycle' - the second part of photosynthesis. Three molecules of carbon dioxide enter the cycle and with the energy of ATP and NADPH create a three carbon compound called PGAL with each round of the cycle. When two molecules of PGAL come together they form a six carbon glucose or sugar molecule that is stored in the plant or fruit. The grand result is that this process causes one molecule of glucose and two oxygen molecules to be produced. The oxygen is released into the atmosphere and the sugar is stored as starch or ends up as fruit. We take the fruit and turn it into wine.

This is the basis of the ecosystem we live in and the oxygen we breathe. Plants produce the oxygen we breathe and can even help deal with all the carbon dioxide which is a greenhouse gas. The lungs of the planet are said to be in the Amazon forest in Brazil. Every minute according to Greenpeace six football fields of Amazon rainforest are destroyed. Some of the plants containing unique compounds which could be used to fight against cancer and other diseases are never to be seen again. They vanish in the smoke of the beef farmers' bonfires. All to make way for beef and soy production. An old Cree Indian proverb says that only when the last tree has died, the last river has been poisoned and the last fish has been caught will man realise he cannot eat money.

With the depletion of the ozone layer, global warming, climate change and in particular radiation, this has caused an increase in grape sugars. This has a knock on effect in wine in that the wines are now being produced with higher resultant alcohol levels of 14% and even above. Most connoisseurs won't drink that wine as the taste will be too hot from the alcohol levels and the flavour and aromas severely impaired.

The Phylloxera Threat

The roots on which grapes are now grown in Europe are all from the American vine. Phylloxera took root and killed many of the vines in Europe. If you have ever wondered why the declaration card you fill in on arrival at an airport is so important, perhaps the story of Phylloxera will help you understand why the importation of plants and animals has to be strictly controlled. Phylloxera's arrival in Europe in the late eighteenth century caused devastation in vineyards right across Europe. People watched vineyards slowly being destroyed by this tiny louse munching through the roots of the European vines. The louse arrived to Europe via a botanical example that was propagated from America when wine makers were exchanging vine samples to help the wine industry. How ironic? The insect spread rapidly in roots with no resistance. The American vine was naturally immune to the attacks but the rest of Europe was not. The wine world had just overcome powdery mildew at that time.

The insect attacked the vine through the roots where it created galls (nodules) and fed on the sap depriving the vine of nutrients and thereby killing it. One has to consider how terrifying this was without the science and technology we now have to discover what the cause was. The entire European wine world was really near the brink of destruction. We need technologies and scientific investigation to be able to guard against this type of thing ever happening again.

After much investigation, the power of which at the time was limited, it was seen that the European vine species *vitus vinifera* should be grafted onto the American vines that impart a resistance. Grafting *vitus vinifera* onto *vitus labrusca* was the only way of protecting the vine and whole vineyards were replanted with this root stock. The problem has re-emerged in recent years with root stock called AXR1 which were developed to be resistant to fungal attacks and other disease. These were planted in New Zealand, South America and California and now they are being replanted as these vines' roots started to be targeted by Phylloxera. You can imagine the costs. Clay soils assist the insect but sandy soils naturally inhibit them which is why you see vines growing in the Bordeaux basin. What would be interesting to see is if a living soil with biochar provided any resistance to the bug? The wine industry is constantly being as vigilant as possible to stop the spread of Phylloxera which is now found sadly in all parts of the wine world.

The only natural European vines that now exist are on the Canary Islands off the coast of Africa. This also reminds us about the dangers of mono crop farming as without biodiversity, crops are more susceptible to disease. Another good example is the scourge of wheat blight. If it takes hold we are looking at famines on a global level. We have to change and we have to fund these science programmes properly to safeguard our ability to produce food for billions of people. Wheat crops are being heavily decimated by this fungus and we need to have scientific solutions and monitoring. The fields have been rust free for over 50 years and now a strain called Ug99 has been spreading via billions of spores.

Grape Science - Colour & Sugar

A grape's varietal determines the wine type, flavour and smell of wine. In reds you will probably know them as Cabernet Sauvignon, Syrah, Merlot and Pinot Noir. In whites, Muscat, Chardonnay, Riesling and Sauvignon. Each grape varietal will have a different wine flavour. Grapes are the fruit of the vine and start with a flower - the plant's reproductive organ. The vine produces flowers to spread its seed. The whole purpose of flowers is to create a fruit which spreads a seed and generates more plants reproducing in the environment. The flower structure contains a stamen, the top of which contains pollen and the base of which contains the ovary. The pollen is sperm and this fertilises the ovary which starts the process of fruit development. In order for the seed to survive, nature in its wisdom packs sugar and nutrients into the fruit in order for the plant to be reproduced ensuring it germinates on its own.

Grapes for wine are smaller than what you get at the supermarket and also sweeter. On average they will have 25% sugar in them which will generate wines of 12.5% alcohol. If you halve the percentage of sugar you get the resultant alcohol level in the final wine. As the grapes are growing the vineyard manager will be walking around testing the grapes and their sweetness in preparation for harvest and wine production. Grapes' skins contain many of their health compounds. A quarter of the grape is made of sugar and the rest is water. The skins contain important aromatic chemicals to form the taste, flavour and aroma of wine.

The anthocyanins give both the colour to wine and health benefits. The tannins give the wine its astringency. The tart taste in wine comes from two acids found in the grapes called tartaric and malic acid. Minerals can also be found in the juice in trace amounts. BSA, or Berry Sensory Analysis, is used to determine the ripeness of the grapes for the ideal harvest and many wineries have adopted it. Fresh grape juice is one of the best known liver tonics and liver regenerators. Grapes in naturopathic medicine and Chinese medicine have a very strong affinity to the liver and its restoration.

Wireless Winemaking & Fermentation Wi-Fi

Winemaking for centuries was done with no added yeasts. Fermentation took place only with the natural yeast that was already present on the skin of the grape. Most grapes have a white powder on the skins and this is the natural yeast. As there is a possibility of spoilage, the area of fermentation science has now over 400 aromatic yeasts to hand to steer the process. When grapes arrive at the winery they are sprayed with sulphur dioxide to kill the natural yeast and any other microorganism that might spoil fermentation. Otherwise if any acetobacter bacteria gets in this will result in vinegar rather than wine overnight. Red wine is fermented using whole grapes, as much of its colour, flavour and texture chemistry comes from the compounds in its skin. White wine is made by crushing the grapes in a press and collecting the juice for fermentation.

After red grapes are picked and de-stemmed they are crushed. They are then put into a macerator that bursts the grape skin and releases the sugars. This is called the 'must' by wine makers. The solid part is called the 'pomice' and this mixture is left to infuse as the natural chemicals from the skins will give the resultant wine its body, taste, flavour and character. Sometimes the grapes are made into a must with the stems if they have browned enough as they will add extra tannins to the wine.

After this stage the fermentation starts and in conventional wine making, the aromatic yeasts are added. There are well over 1000 different types of yeast in nature but the one for wine is Saccharomyces cerevisiae and it outperforms all the other yeasts and can tolerate high levels of acid and alcohol. The yeast's desire to dominate through evolution found a friend in winemakers who wanted them to produce alcohol which other yeasts are unable to tolerate. The grape juice is now very rich in sugar which is a dream environment from microbes. Yeast is the traditional one when it comes to the production of alcohol. Bacteria divide in two but yeasts bud multiplying in a very short period of time.

If a bacterium divides every 15 minutes it's going to take a lot of time for the bacteria to grow and for the fermentation to take off. This can cause a hijack by other microbes which will spoil the juice. Yeasts can bud infinitely so the inoculum will grow quickly, eventually boiling. If the right yeast takes root then wine will be produced. If not, the liquid will spoil, hence the intervention with aromatic yeasts to ensure the right fermentation takes place from the outset. Yeast takes the sugar in the grape juice and produces alcohol and carbon dioxide. When the yeast converts most of the sugar to alcohol in the wine, both the limitation of food and yeast sensitivity to the alcohol naturally stops the fermentation.

The primary fermentation in wine generally lasts for between seven to fourteen days. When the fermentation is complete, whereby the sugar in the wine has been converted to alcohol, the level of alcohol is 12% on average. With sugars in the fruit at a point of generally 25% winemakers got 12% levels of alcohol. Due to climate change and global warming there is more photosynthesis in the plant which produces more sugar in the fruit resulting in higher alcohol wines of 14%.

When the wine has finished the primary fermentation, it's time for the secondary fermentation - the conversion of acids. Bacteria is used to convert malic acid to lactic acid. Malic acid is found in ripe green apples and is very good for breaking down bile stones naturally. The lactic acid which is found a lot in dairy products gives wine a buttery effect. This is also known as malolactic fermentation and as the name suggests, the conversion is from malic to lactic acid. The fizz of Champagne is a result of this process and takes place within the bottle as it ages when it's racked. A must that is allowed to naturally start to ferment takes a bit of time to begin this process which is why big commercial operations use aromatic yeast to kick-start it. This guards against spoilage and the huge risk in obtaining style in the resulting wine, just trusting nature. However winemakers will agree that a natural ferment with the yeast on the skins of the grapes will produce higher levels of glycerol which are said to have a fuller rounder palate structure in terms of sweetness and smoothness. The fermentation lag phase of *vin au natural* also allows more anthocyanins and phenols into the wine which helps colour stability in red wines.

Apart from the ethanol and CO_2 that the yeast metabolism produces, it also generates glycerol which helps the body of wine. As the fermentation increases, the alcohol water mixture also helps in terms of further extracting chemicals from the grape skin in the liquor and in red wine this is especially important. Let's not forget that alcohol is a very powerful solvent. The yeast themselves, acting as 'biotransformers', transform various compounds into taste and aroma molecules. Wine Tweaking? Well, to help wine come to a good quality, it may be necessary to use some additives. If it's not organic wine that is. The wine may need tartaric acid to help the acidity, clarifying agents, enzymes and preservatives such as sulphites which have been much in the media. This is why organic wine is so popular - it has very low levels of sulphites which cause allergic reactions in some people.

'Racking' is the next step and after the fermentation is finished this takes place. Wine is taken from the vessel leaving the plant material, skins and seeds behind. The wine is passed through a filter to remove any remaining solids. 'Fining' is when either clay or egg albumin is added at this point to help clarification of the wine. Both these substances attach to any clusters of yeast in the wine and fall to the bottom of the vessel. The resultant wine is then either placed in barrels to age or is bottled. If it's put in barrels, the length of time for ageing is a minimum of nine months to years depending on the prestige of the wine. White wine goes from stainless steel tanks to the bottle, except some Chardonnays which spend time in a barrel to give them oak flavours.

Some very big donations have been made recently to wine science centres by high net worth individuals wanting to leave a legacy. TJ Rogers, a millionaire from the semiconductor business, is also an avid wine aficionado and makes Clos De La Tech. A fine Pinot. He is the founder of Cypress Semiconductors which has $1 billion of sales a year. He has foot the bill with a $1 million donation to the new research centre.[6] This donation funded the purchase of 152 tanks that can hold 200 litres each of wine. Rogers who also has his own wine operation wants to show what science can do for the wine industry. Within biotechnology fermentation has to have controls.

Adding sensors and modelling what happens mathematically in terms of sugar, alcohol and heat, you can better understand how long to leave the must to infuse, when to add the yeast optimally and what affect heat has at every stage of winemaking. You get the info in real time. The electronic devices have been made by Cypress Semiconductors and are all wireless. They relay information every 15 minutes in a loop creating the world's first wireless fermentation system. Rodgers wanted to produce a new range of fermenters based on his home prototype that could revolutionise wine making. The professors agreed at a meeting about his winery where he showed them his new piece of high tech kit. Many wine families are giving back to the wine industry through philanthropy.

The new wine research facility at UCDavis also aims to be carbon neutral and eco-friendly.

A device that could also become very popular with the new wave of wine makers is called a WinePod and will cost you $5,000. This high tech pod looking very futuristic is a home winery. This is an all-in-one system that presses, ferments, monitors temperatures and can sit in your kitchen or garage being only four feet high. The pod is connected to software on your PC that assists you to start making wine. This takes your wine making passion to the next level. The WinePod website also has an online school with video presentations of all you need to know for wine making. The company started with funding of $4 million. All the info is online especially the wine making areas. This system could encourage so many people to start making wine again. They are currently looking for more funding to expand.

Oak Casks & Flavour Chips

Most red wines end up in oak barrels to age post-fermentation. Some whites do but most are made in stainless steel tanks and have a fruitiness that the oak ageing will clash with. White wine is only put in barrels to give the wine more creamy flavours. Placing red wine in oak barrels to age adds secondary aromatic compounds and flavours. A new barrel will have 30% or 50% new oak and the rest of the barrel is made up of neutral oak. The more new oak the more flavour. The wine is left to age in the barrel from 8-12 months. When the wine in the barrel has reached an oak saturation point it's moved to a neutral barrel to further age and soften up. A neutral barrel is one that has been used for three vintages and there is no further aromatic compounds left in the wood. The elegance of the French barrels is great for Pinot and Syrah. The spice of the American barrels and the resin works well with Zinfandels.

There are three types of oak barrel generally: French, which gives an elegant soft oak flavour, American, which has vanilla flavours and soft Eastern European wine barrels. The barrels allow the wine to breathe and release water and alcohol and concentrate the flavour and structure of the wine. The oak planks that make up the barrel by the cooper are passed through a fire of oak chips and toasted. This resultant 'toast' can be soft, medium and hard in terms of taste imparted to wine enhancing woody flavours. Oak chips are used by some wineries to get flavour into the wine in matter of weeks what would normally take an oak barrel over a year to achieve.

Many American Italians are becoming interested in winemaking again as many of their grandparents were home winemakers. As more and more people get interested in wine, especially Millennials, they will want to start making their own quality wines on a small scale. The excitement, indeed challenge, will be too much to resist getting their hands wet.

4

Wine Transfusion

In 1991 '60 Minutes' changed the wine industry forever and grew the wine market in the USA by 44% and elsewhere exponentially. The scientific connection between wine and health was made public through the 'French Paradox'. So why do the French, who eat lots of fat and drink wine not have heart disease?

As we have seen through previous chapters, historically wine has always in some way been connected to health, medicine, wellbeing and sometimes even mysticism. There has been wine as a safe way to ingest liquid, medicinal uses when herbs were added, the tradition of French physicians prescribing wines and now through '60 Minutes', a scientific basis beginning to emerge between wine and health. The scientific case between wine and health was flung open, showing wine to be a natural preventative medicine for diseases such as heart disease, cancer and diabetes and making a strong case to refer to it as a health food. With modern science we know that the compounds in wine that increase health are B vitamins, minerals, resveratrol, bioflavonoids (Vitamin P) and polyphenols.

There are over 1,000 phenolic chemical compounds in wine and so far we know that two of them, resveratrol and proanthocyanins, have significant health benefits. They come from the skin of red grapes. This is why maceration of the grapes and letting them infuse in the must for hours at the start of winemaking is so important in terms of the health benefits that the wine will impart. Over 52% of hospitals in the top 65 metropolitan areas of the US offer a wine service to their patients.[1]

60 Minutes That Changed Wine

Actually five minutes. When Morley Safer from '60 Minutes' did a segment on The French Paradox, no-one could have predicted the effect those five minutes were going to have on wine and health.[2] Remember these were the days a few years before Dean Ornish's ground-breaking paper linking diet to heart disease. This paper was the first that really made a case in the public domain about why diet was so important for wellbeing.[3] Professor Serge Renaud is the father of The French Paradox and was a Professor of Nutrition. He grew up in Bordeaux with grandparents who owned a vineyard.

The first physician in fact to make a connection with The French Paradox was an Irish pioneering cardiologist Dr Samuel Black who is known to have been the first to have described angina pectoris. The stressed, obese and overeating, he found, were more prone to the disease. In 1795 he published what he saw from angina pectoris post mortems. "On examining the coronary arteries, I found, with a mixture of satisfaction and surprise, that they were completely ossified through their whole extent." He also did further research in diabetes. Black in 1819 went to the south of France and saw no examples of heart disease there, about which he wrote, "the French habits and modes of living, coinciding with the benignity of their climate and the peculiar character of their moral affections (psychological stress)." He wanted to know why there was such a disparity between the Irish and French in terms of heart disease in that the French seemed not to suffer it.[4] So there is even some historical evidence to suggest the French Paradox even before the industrial revolution and modern farming.

The Copenhagen Heart Study suggests that there is a connection between moderate wine drinking and a reduction in heart disease and longer life expectancy. This study which spanned the years 1976 to 2003 and contained over 13,000 participants was published in The British Medical Journal.[5] The study is important as Denmark is not a Mediterranean country so it does not have that diet or lifestyle. Many thought the wine benefit was only in countries that were in the Mediterranean. The scientists also attribute the reduction of heart disease in the country by 30% to increased numbers of wine drinkers since they entered the EU.

The '60 Minutes' piece opened in Lyon talking to a Dr Curt Elleson, a cardiologist from Boston who wanted to find out the biochemistry behind the French Paradox. In a rich and developed country like the USA this was a time also when fat was bad. We know more about fat in the diet now and if one food is more dangerous than any other - it's sugar. You just need to read New York Times bestseller 'In Defense Of Food' to see why. Fat at that time was being held responsible for all dietary related problems. So when this paradox emerged which highlighted that despite the French eating lots of fat in their diet, they were not getting fat, naturally eyebrows were raised. Most importantly of all they did not suffer coronary heart disease. The reason, scientists believed was due to the inclusion of wine in the French diet.

Since the first airing of the show a huge debate has been taking place over the last two decades about whether or not the paradox can be explained by wine. Physicians and specialist cardiologists started studying wine as a result of this broadcast. The debunkers have said the population statistics are the same and they can't see why wine would be an effective agent when the French drink less than some of their European neighbours. Since that time and perhaps thanks to that programme, more scientific medical research has been done in exploring the connections between wine and health in a variety of different diseases and has shown wine to be very beneficial. Even Americans visiting France have been shown to lose weight even if they don't drink wine.[6] Although the French eat a lot of saturated fat they don't eat anywhere near the level of trans fats that the Americans do. If you're eating trans fats your risk of developing heart disease and diabetes is very high.

This was also the age of coronary bypass surgery and fatal heart attacks and all you needed to prevent them was some wine in your diet from what the studies show. This also caused a revolution in the health food industry whereby diet was being taken very seriously in the role of disease induction which we now know that diet is the cause of. Nevertheless before that not much research was being done nor was the role diet played in health well understood. This was just before the 'gene rush', the discovery of the polymerase chain reaction and before we even had a sniff at a stem cell. So you can imagine how important this was at the time which is why it caught peoples' attention. Supplements were being sold but mainly high doses of ascorbic acid sold as vitamin C which is not natural vitamin C. You're better off drinking lemon juice as a source of vitamin C rather than taking ascorbic acid pills.

If you feel the scientific debate is confusing just look at the population studies and the faces of the wine drinkers. Two examples of how people can get confused by reports in the media undermining natural health are soy and organic food. The fact of the matter is that women in Asia, particularly in China, who eat soy products reduce their risk of breast cancer.[7] In fact in Japan they don't even have a word for menopause as we do in the West as they have very low levels of this condition. However when Chinese women move to the USA, in a matter of three years, their risk of breast cancer increses.[8] One possible cause is that they adopt some of the dietary habits of high calorific intake. Recently there was also an attempt to defeat the organic food movement with attempts in the UK to say that organic produce was just the same as conventional vegetable produce. Nice try but that study was flawed and designed in a way to undermine the reality. Ask anybody who eats organic food and they will tell you how the food differs especially in terms of mineral content and lack of pesticides. The study from the Soil Association was far more robust and credible containing over 400 peer reviewed scientific papers.

So when people start to try and debunk wine's health benefits stick to the scientific facts you have learned in this book and the large population studies. Large population studies matter to you. Those facts speak for themselves and the millions of people down through the centuries who have benefited from wine. Especially the longevity and health of the French and the people of Sardinia which is attributed to wine.

The History of Wine & Medicine

Wine since ancient times has been seen as a panacea of health. In Mesopotamia and Egypt there is archaeological evidence that wine was viewed as a medicine. In the Vedas of India it is synonymous with Soma the god of healing. In Greece Hippocrates, the father of modern medicine wrote it into his pharmacopeia. In China the Emperor Yu found he was cured of insomnia from wine. Asclepiades, a Greek doctor, seems to be the first we know of who started prescribing wine for a whole raft of ailments in Rome at around 124 B.C. He was very successful and prescribed wine containing herbs for a whole host of diseases. His nickname was 'the wine giver'. As medicine strolled through history it held the hand of wine.

Wine was seen as a health tonic and restored blood, giving a boost of energy and lifting the spirits when there was fatigue or depression. Christian monks in France had a long tradition of wine as a digestive to keep the monks in top health. Hippocrates, whose medical oath physicians still take to this day, "First do no harm", saw that wine helped the four humours of man. Paracelsus, the alchemist of the middle ages, saw wine working in much the same way as we do adaptogenic herbs today. These help assist the biochemistry of the body in times of stress.

The Doctrine of Signatures which looks to a particular mark or colouring in a plant when seeking to discover what uses that plant may have in the treatment of people, regarded red wine for the blood and white wine as drinkable gold, or in Latin 'aurum potabile' as it has absorbed the energy of sunlight. The medical school at Salerno which no longer stands was the first medical school in the West and for at least three hundred years was a hub of activity. This medical epicentre had medical texts from Asia translated from Arabic. Wounds were traditionally treated with poultices of red wine and today we know that the wines of the Medoc have antibiotic properties. Even Gladiators who had their stomachs cut open in battle with their entrails hanging out, were washed in wine and were sewn up. Many of them survived due to the antiseptic qualities of wine.

One other use of wine was to combine it with medicinal herbs, as the alcohol acted as a great solvent and also preservative. When you consider that the cost of spices at that time was equivalent to and as precious as gold, you can see why wine catalysed medicine. We know today, through the use of scientific tools, that many spices have very powerful effects on the body. Clove is the most powerful anti-oxidant according to the ORAC (Oxygen Radical Absorbance Capacity) index and rosemary follows. Ginger acts as an anti-inflammatory. Cardamom is useful as a tonic for the stomach. Fennel helps the digestion. So wines infused with these herbs and spices, with alcohol working as a solvent would have had higher levels of the active principals and would have stored well. They would have had to as they were very expensive. In China these herbal wines were called Spring Wines for their tonic properties.

Archaeologists recently did studies on artefacts found in the crypt of an Egyptian Pharaoh and found that the wine urns had traces of numerous herbs.[9] At the start of the 16th century wine was known as a therapeutic agent in the pharmacopoeias of the major European countries. Then the wine world and health had a big drama. The dramatis personae were a King, his physician with the title of *Premier Medicin*, the King's mistress Madame de Maintenon and a rival doctor from England. Who had the 'healthiest' wine was the *dénouement*.

Physicians Du Vin - Wine Prescribing Doctors

All Royal courts in the middle ages had a physician, and more than likely one who was very talented. One who was not going to poison you with potions as mercury was doing the rounds at the time. In terms of Tibetan Medicine that I am also concerned with, every Chinese emperor had a

Tibetan Doctor in the household. Even the Russian royals did such was their renown and abilities. If you were a royal you needed a great doctor. France's King Louis XIV was no different. His personal physician was Antoine d'Aquin and he had the King drinking Champagne at the palace of Versailles as a preventative medicine daily with his meals. D'Aquin was trained in medicine from the University of Montpelier which had become the new Salerno. He attained the position of *Premier Medicin*; not only was he responsible for the health of the Sun King but had total power over France in terms of health as a result. Tibetans say a doctor's position is a powerful one, in that, even a King has to listen to his advice. As the King aged, his mistress Madame de Maintenon preferred a new rising star in the Paris medicine scene – a doctor from the UK called Guy Crescent Fagon who was her *protégé*. As the King grew older she knew she would have more control by owning the doctor who was very powerful in royal court life and hierarchy. Fagon however preferred Burgundy wine and stopped all the Champagne.[10]

Due to his position as *Premier Medicin* there was a huge reaction. Today it's a bit like a celebrity endorsement to Burgundy for its healthiness over Champagne. The result of this move caused both sets of producers to jump into historic marketing wars. The Champagne lobby paid medical students in Paris to write a thesis each about why their wines were healthier and printed pamphlets and distributed them. Burgundy at this point, cool as the cat that had swallowed the canary, just got the Dean of Beaune Medical School to come to Paris and give a public talk to the influencers of Parisian society at the Faculty of Medicine. Excellent strategy as this was long before public relations as we now know it. The competition for which was the healthiest wine went on for a further 130 years.

By the 1850's as many as 175 wines were noted in the German pharmacopeia for their health benefits. There is a clear connection to a long tradition between wine drinking and health in France. We know today that they were benefiting from wine, as modern science has proven the health benefits they believed were in wine with many other herbs. Many pharmaceutical drugs are based on the same herbs. Then along came prohibition in 1918 in the United States and all wine was banned. The interesting point here is that this happened at the birth of the pharmaceutical industry. Their biggest competitor was wine and all the wines fortified with herbs. Plus homeopathy - one in six physicians was a homeopath.

For nearly 25 years wine was banned and therefore the connection between wine and medicine was set adrift. Interestingly, a generation of American physicians grew up in this era? The purpose of this book is to encourage them and their patients to return to wine as a daily health regimen.

The new generation of doctors qualifying are very taken up with preventative and alternative medicine. They understand the power of natural medicine.

Traditional Medicine for Wellbeing

Of all the people who have written on this subject, Dr E A Maury has written the most. He was the husband of the founder of aromatherapy, Marguerite Maury, a biochemist who became as equally renowned for her impact on natural medicine. Dr Maury was a French physician who was also qualified in acupuncture and homeopathy.[11] When it comes to wine Maury believes that wine acts as an aid to digestion. Medoc wines are great at toning the stomach with their tannins, calcium and magnesium. Also the intestine walls get toned, particularly if the person is prone to diarrhoea or irritable bowel syndrome. For constipation, he recommends sweet wines like Anjou which are high in glycerol and which will have a mild Epsom salt effect. Another promoter of wine and health was Dr Luca California. He also wrote a number of books on the subject. Since their books were written, we have way more scientific studies to prove what they were saying at the time about wine and health. We have seen the same in herbal medicine whereby for centuries herbs were used to treat disease. With the aid of modern scientific medical investigation many of the healing powers of plants have been proven.

Did you know 70% of heart attacks are fatal? About half of us will die of heart disease. Most recently, Professor Roger Corder who wrote 'The Red Wine Diet', has found very strong connections between wine and heart disease prevention. He gives very strong scientific evidence to back up what Renaud believed and investigated. Professor Corder is based at the prestigious William Harvey Research Institute in London and has investigated the relationship between wine and health for over 15 years. Corder focused on a particular area in the south of France that he and his team believe is the key to understanding the wine and health phenomena. He wrote his bestselling book 'The Red Wine Diet' based on his findings.

This area of France more than any other had the longest life expectancy. He says that in 1970 the lowest deaths were due to the highest consumption of wine. Corder stresses the point that heart disease is a silent disease and that it progresses in a way you are not aware of until either a stroke or sudden death. Interesting to note here is that 70% of heart attacks are fatal. Over time, fatty deposits can build up on the walls of your heart vessels eventually becoming blocked. As a scientist he wanted to see what the protective components in red wine were. Corder found that red wine does decrease the damage to the wall of the blood vessel called the endothelium. His research group found the compounds responsible for this effect - polyphenols, in particular procyanidins. He then started to explore two regional areas - Sardinia and France.

Sardinia is known throughout the world to have a healthy long-living population. He found in the south west of France many men living over the age of 75. In the North, he found the beer and white wine drinkers had the lowest longevity rates. A place called Gers near Toulouse in the south had the highest rate of all and Corder found that wines in the area are made with the Tannat grape. This grape has a very high tannin content and also procyanidins at levels four times more than most other wines. So if you want to protect your heart, you will need a low alcohol level wine which will have a higher content of those proanthocyanins. By the way Tannat grape wines go well with a steak. So perhaps steak houses could start serving it on the menu which would be good for consumers' health and also producers.

The Health Compounds

The active compounds of wine for health are B vitamins, minerals and phenolic compounds, such as anthocyanins, catechins, quercetin and resveratrol. The B vitamins are Vitamin B1 (thiamine), Vitamin B2 (riboflavin) and vitamin B3 (niacin) specifically. B vitamins are essential for energy production in the body. Vitamin B1 is good for the pumping strength of the heart and levels are depleted with long term use of diuretics. Thiamine is also good for the nervous system and protects the nervous system from damage particularly in diabetic neuropathy. Plus it improves your thinking and was the first vitamin to be discovered. Vitamin B2 helps in the production of thyroxine hormone which controls metabolism and red blood cells when it works with iron. The main benefit of riboflavin is that it protects the eyes particularly the lens against cataracts. Half of the body's requirements for Vitamin B3 come from the amino acid tryptophan. Being a natural anti-inflammatory it has benefited those with arthritis. In nerve health it is said to ease depression, anxiety and insomnia.

Some would say that the levels of the vitamins are very low in wine and they are right. However the vitamins found in wine are natural vitamins and not 'chemical vitamin' commonly found in supplements. Natural vitamins contain a lipid, carbohydrate and most importantly a protein. The protein within our biochemistry acts like an address on the nutrients, without which the compounds won't get delivered. Therefore it is more effective, utilised and retained than its chemical cousin which today is being produced by the ton load in China. Some people prefer natural vitamins. Interesting info for those who take supplements daily - one in three people in the western world.

Two times Nobel Prize winner Linus Pauling said, "Every sickness, every disease can be traced to a mineral deficiency." Wine contains a number of minerals important for health: calcium, magnesium, phosphorous, potassium, iron and selenium. Calcium is needed for bones, phosphorous for the biochemistry to produce energy, potassium regulates heart beat and blood pressure, magnesium is involved in over 300 enzyme related processes in the body and selenium, apart from being an antioxidant, has a huge role in fighting cancer. The levels of minerals found in wine are down to the levels found in the soil. Vine roots as a result of modern farming only grow to a depth of a meter and grow sideways searching for food. Vines grown biodynamically remember, grow to a depth of 20 meters. Minerals that humans need are phosphorylated and the plants do this for us. That is where we get our minerals from and why organic food has a higher mineral content as it is grown in mineral rich soil.

As the top soils rich in minerals have been eroded, conventional vegetables have lower mineral content whilst containing pesticides throughout. Not just on the skin as many people believe. A study to compare the mineral content of different wines, especially organic and biodynamic against conventional, would be very interesting. Once again mineral supplements from your pharmacy can be best described as ground up rocks. Your body needs minerals in a food form not from a chemical factory. The more you source your minerals from nature the better for your health and wellbeing. Remember in the last chapter when we spoke about chlorophyll the green pigment of plants which is a natural protein? This has exactly the same structure of the blood protein haemoglobin; the only difference is that the mineral at the centre is iron rather than magnesium. Port wine is particularly high in iron content and has been used in the treatment of anaemia historically. Anaemia is a condition where there are not enough red blood cells. Seeing that iron is a central building block of red blood cells you can see why port has helped people and gained renown in the treatment of anaemia.

Wine in Moderation Campaigns

Wine moderation campaigns have recently been springing up and when you look at the facts you may wonder why. Why on earth does there need to be a moderation campaign on wine drinking which is naturally a culture of moderation? If anything we need to be getting more people drinking wine especially heart patients. True, a small proportion of people drink too much wine but they drink too much of anything.

Why should wine drinkers suffer? They are buying cheap cider not fifteen dollar wines to have with food. Two researchers, Dr Klein and Dr Pitman, found that wine drinkers in comparison to all other forms of alcohol drinkers drink very moderately. They also found the majority, over 80% of wine drinkers in the U.S., drink their wine at meal times. These findings were published in the peer reviewed 'Journal of Substance Abuse.'[12]

Additionally the campaigns are making wine drinking taboo. Subliminally speaking that is. They use words like 'moderation' in their campaigns and this is sending out a sense of shame. Wine drinkers are moderate drinkers and are healthier and if anything they should be applauded. Are the same campaigns ever seen on whiskey, beer and even more importantly the notorious alcopops? Nope. In addition to get a medical angle to these moderation campaigns, they include 'fatty liver' which is a huge problem in medicine today and especially with alcoholics. However apart from alcoholic induced fatty liver there is just as much, if not more, non-alcoholic fatty liver called medically 'NASH'. The cause of this is metabolic syndrome mostly and sugar.

Do you ever hear a moderation campaign about sugar particularly high fructose corn syrup? Take wine away and you can say hello to obesity and a whole host of other diseases that are far worse for society. Ones that they are saying wine causes ironically. What about the dangers of fizzy sugar drinks? They are actually far more dangerous in terms of health risk. People drink glasses of wine but litres of fizzy sugar drinks. Some people in the USA drink litres of the stuff a day acidifying their biochemistry which works better at a slightly alkaline level. What about the dangers of the phosphorous in colas which leech the calcium from your bones causing you to need 15 litres of water to bring your blood pH back to normal afterwards?

If you don't believe me ask the Belgian school girls who were hospitalised due to excessive cola consumption brought about by a marketing promotion.[13] Over 100 were poisoned. What is the EU doing about this? Are they protecting children from colas with no nutritional value?

These moderation campaigns if anything should be saying two glasses a day will keep the doctor away. And more importantly a heart attack. They should be encouraging more people to drink wine for the health benefits. Getting the two glasses a day message out there would also get their moderation message to the public without shaming wine drinkers. If anything the money and ethos of these campaigns should be spent saving lives and to do that you have to put the message of two glasses of wine a day in the vernacular. Moderation in a consumer world psychology means bad or naughty or dangerous in the unconscious. No. No. No. We need to drink two glasses a day for wellbeing, health and longevity. Wine drinkers probably save a national health budget quite a lot of money, billions in terms of preventative medicine and this needs to be researched and appreciated.

Both Russia and China, two world powers, have each been communicating that they are encouraging their populace to drink red wine rather than spirits. In Russia, vodka is the main cause of widespread alcoholism and mental health issues. All the research points to two glasses a day max and if you go over that level, wine starts to become damaging. Paracelsus himself said, "only the dose permits something not to be poisonous." Jefferson also hoped the USA would become a wine drinking country which is why he spent so much time studying in the European wine worlds. He believed that wine would be a culture of moderation and would bring stability to the New Republic that spirits would not.

So drink wine to the betterment of your health and ignore these moderation campaigns which are not relevant to wine culture and you statistically speaking. Enjoy your two glasses. Savour them with your family and friends knowing they could be saving and extending your life.

And spread the word. Ignore the 'nanny state'.

Symposium - Social Effects of Wine

Wine is a social lubricant. The word symposium comes from ancient Greek which means to drink wine together. The occasion was for men of noble families to come together, drink wine, relax and discuss philosophy, politics, religion and poetry. This is the natural setting for wine and what it can induce. One of the biggest functions wine has, people say, is that contributes to their relaxation and enjoyment of life. In 375 BC Eubulus the playwright has Dionysus, the God of wine say:

"For sensible men I prepare only three kraters: one for health (which they drink first), the second for love and pleasure, and the third for sleep. After the third one is drained, wise men go home. The fourth krater is not mine any more - it belongs to bad behaviour; the fifth is for shouting; the sixth is for rudeness and insults; the seventh is for fights; the eighth is for breaking the furniture; the ninth is for depression; the tenth is for madness and unconsciousness."[14]

In recent years studies have shown that what makes people happy is a social group or lots of friends. All of psychology can be reduced to stimulus and response. So be aware of what your primary stimuli are as you will be the response. For those who believe money makes you happy it can't buy you excitement.[15] A UK study showed that when people had more than a few thousand pounds in the bank their happiness feeling started to plateau. When wealthy people were asked what they valued most they said their friendships. People gathering together and relaxing with wine and food is a rich lifestyle. This is living and is a way to do what bioenergetic psychotherapists call 'discharge'. In other words let your coils all unwind. Wine is the subject of a huge amount of social media conversations.

Quality of life and particularly meaning, has become a very important topic in recent years. The debate has grown into an awareness of what 'spiritual intelligence' or SQ is. You have probably heard of IQ and EQ. This lack of meaning has also been spoken of as 'Dionysian starvation' a lack of joy in one's life that the God of wine Dionysus could heal. People today live to work rather than work to live. There has to be balance and time for life. Time for joy and celebration with family and community. In the end what does your dollar mean if you can't enjoy it or create a quality of life with excitement?

Communities in the west have been breaking down. People think it's cool to condescend this is their main pastime. Narcissism is quite rife so much so now that we have what psychiatrists call 'the garden variety psychopath' in increasing numbers. Charming liars. By significant contrast Mother Teresa talked about the poverty of the West being a lack of love, "Being unwanted, unloved, uncared for, forgotten by everybody." Wine can bring people together naturally to share life and have fun. This will increase your happiness factor. There has been a resurgence of community. Social entrepreneurs abound. Co-ops are being created everywhere. We are social beings and always will be.

The compounds wine contains give you protection from a whole host of diseases which is the subject of the next chapter. One of which has been connected to switching the gene for anti-ageing on. The new medical 'magic bullet'.

5

Wine as Preventative Medicine

In 2003 a scientific paper was published in 'Nature' that connected a compound in wine called resveratrol to extending the lifespan in yeast by 70%.[1] The group went on to synthetically make the compound through a biotech start-up Sirtris. They sold this company, whose technology is based on resveratrol for $720 million to Glaxo-Smithkline.[2] This compound is one of 1,000 potential health compounds that wine contains.

When David Sinclair saw that the compound that was most powerful in extending the lifespan of yeast was resveratrol, he nearly fell off his chair. Sinclair was an Australian biologist who came to Harvard medical school to study ageing. At the time he was studying the anti-ageing genes called sirtuin and was keen to see what switched them on to cause longevity. Sinclair had gone through studies with many different compounds only to discover that the activator of the enzyme siRT1 was none other than resveratrol - a chemical compound found in wine. He published his paper in 'Nature' and this was spotted by a savvy biotech entrepreneur and venture capitalist come catalyst, who also happened to be a Harvard MD and PhD called Christoph Westphal.[3]

Resveratrol & the Venture Catalyst

Westphal was an entrepreneur at heart despite being a physician. He knew his real vocation was to start biotech companies that spun basic research into drugs that worked. Drugs that treated disease rather than managed a condition. He met with Sinclair after he saw the scientific paper and applied his Midas touch to the project and set up Sirtris in 2005. They managed to raise over $80 million for the company with Westphal using his 'venture catalyst' skills. The pair had a strategy first to build on what Sinclair had discovered. Westphal went and found Sinclair which is also a unique point in terms of the story and the world of biotechnology venture capital.

The sirtuin genes when switched on created the benefits of calorific restriction by generating the sirtuin enzymes. This happens all over life when the organism is under stress. The calorific restriction slows down the ageing process and protects against cardiovascular diseases. A bit like what may have happened to the seniors in the movie Cocoon. Remember that? They accidently found the fountain of youth and started to grow young again. The sirtuin enzymes are proteins, biological catalysts that go into the mitochondria (the engines) in the cell and renew them. A bit like replacing an old car engine with a new one. Resveratrol, a compound in wine, was one of 20 compounds that they found would enhance the activity of this enzyme in the body. They figured out that to enhance the enzyme activity by using levels found naturally in wine would require a person to drink a 1,000 bottles of wine a day. Therefore they created a chemical lab form and started seeing what would happen in mice. They fed two groups a high fat diet, one with their resveratrol compound and another without.

What they saw was that the mice without resveratrol had the typical physical effects of a high fat diet such as fat seen on human organs. When they looked at the resveratrol fed mice they found the organs were healthy despite the high fat diet and actually did not gain that much weight either. siRT1 is the main target of the resveratrol Westphal et al have created using biotechnology. The mice ran faster in studies and also did not have fatty deposits despite their high calorific diet which is also called 'the western diet'. Their muscle tissue was very lean and full of mitochondria which are the engines of biological cells. Their insulin sensitivity had also increased which is very good news for 100 million Americans with the insulin resistance condition which can lead to a heart attack.

Using a compound found in wine they have attempted to tackle diseases of ageing such as diabetes, cancer and heart disease. They need to do many more trials but initial results are promising, so much so, they sold the company Sirtris. Sirtuins biochemical effects are now being studied all over the world. They hope to reverse diseases with what they have and to design a drug which is currently being studied in North America and Russia to combat type 2 diabetes. They have shown that it does indeed reduce blood sugar levels. If it succeeds it could change medicine as insulin is a multi-target hormone. The only side effects of the drug is that it will also be multi target in that it will stop so many ageing processes and will enhance other parts of the biochemistry. In some ways it's as important for humanity as stem cells. They could change medicine with a compound based on one naturally occurring in wine - resveratrol.[4]

The resveratrol compound is also believed to have anti-cancer properties and human trials are beginning with other research groups. Resveratrol will be examined more for use in treating Alzheimer's as in mice fed the compound it caused large reductions in brain plaque. Exciting developments ahead.

Cabernet Heart Tonic

One of the main health reasons for wine drinking for the last two decades was wine afforded heart protection. Wine contains chemical compounds, not just for the health of the heart but for the entire vascular system or blood vessels. The total length of your vascular system is 60,000 miles. As we age and wear and tear starts to set in, plaques will start to build up. This is a hardening of the arteries. The most serious plaque is in the heart which will eventually cause a heart attack. The plaque starts to build as a result of damage to the blood vessel wall and also through metabolic problems like insulin resistance.

The plaque acts like a plaster cast to the damaged area. One reason the wall gets damaged is due to diet, free radical oxidative damage and lack of vitamin C with bioflavonoids. Another big vascular problem is intermittent claudication which is a cramping in the calf muscle that comes from atherosclerosis or plaque build-up in the blood vessel walls. After two to five minutes walking a person will start to feel cramp in the back of the leg in the calf muscle. If people start to have chest pain that shows heart disease, they may very well end up doing heart bypass surgery which could be possibly avoided by drinking wine. Furry arteries full of plaque reduce the amount of blood and nutrients that travels to the muscle. Wine is a superfood for the heart and vascular system. Wine has been shown to lower bad cholesterol LDL and increase good cholesterol HDL making blood platelets less sticky.[5]

Wine contains bioflavonoids (vitamin P), resveratrol and proanthocyanins. Bioflavonoids act as anti-oxidants for free radicals that damage the vascular wall. In Trinity College Dublin medics have shown that Vitamin C reduces polyp formation in the colon.[6] Polyps can lead to colon cancer. Vitamin P are vitamin C co-factors that make vitamin C more effective. The free radicals are created naturally from the biochemistry but if we don't have anti-oxidants then they can cause lots of damage and stress to the body. By taking anti-oxidants in your diet, you are decreasing the damage to the blood vessels which will cause the plaque of arteriosclerosis to build up in the first place. Resveratrol, as we have seen, acts on sirtuin enzymes which generate the effects of calorific restriction and being younger biochemically. In heart health, it acts as an antioxidant to LDL cholesterol and reduces blood clot formation that could lead to a stroke. Proanthocyanins focus on repressing endothelin-1 which is a protein that constricts the blood vessel and tones the elastin and collagen in the vessel wall thereby protecting it from damage.[7]

The minerals magnesium and phosphorous also help the heart and vascular health and are found in wine. Over 90% of people are magnesium deficient and the mineral also guards against allergic reactions. Studies of people who had high levels of magnesium in their water showed their risk declined by 19% of a heart attack.[8] Lowering blood pressure, inhibiting blood clots and widening the arteries are more benefits. Potassium can reduce a risk of fatal stroke by 40%.[9]

The Copenhagen study showed that those who had moderate amounts of wine in their diets reduced their chance of a heart attack by 50%.

Anthocyanins are found in the skin of grapes and give wine its colour. They are a natural plant sunscreen and act as an antioxidant, sweeping up damaging free radicals in the body. Catechins, also called procyanidins, inhibit endothelial cell production of endothelin-1 (ET-1), a highly potent vasoconstrictor which contributes to heart disease.[10] However, they are only found in adequate amounts in wine from the south of France and Sardinia. Quercetin reduces LDL cholesterol which is the bad one and also inhibits the growth of cancer cells. The compound also reduces allergies. Magnesium too reduces allergies. A Dutch study conducted by Dr. Michael Hertog showed that consuming 16-30 mg of quercetin a day has been shown to reduce heart disease in men by 50%.[11]

By far the most popular and most known wine compound is resveratrol. The discovery that this compound had anti-ageing properties was another major landmark in the wine and health popular science story. Resveratrol has also been shown to stop blood platelets sticking together which will if they stick, cause a heart attack. We now know it also has anti-inflammatory properties which can help wellbeing as many diseases are inflammatory. Resveratrol may reveal may reveal more benefits as research continues in many universities.

Polyphenol Insulin Effect

Due to a high calorific diet now prevalent in the western world there is a very large incidence of diabetes, mainly type 2. Over 6% of Americans now suffer diabetes. This is the non-insulin dependent type. Even more important than this fact, is that over 100 million Americans at least, have metabolic syndrome or what is called pre-diabetes. Over 90% of diabetes is type 2 and is due to obesity. Metabolic Syndrome or Syndrome X is diagnosed for a person with high body fat, elevated tri-glycerides, high LDL cholesterol and a big waist line.

The Nurses Health Study that started in 1976, and has monitored the health of over 120,000 nurses, also has seen that people who include moderate wine drinking in their diets both live and age longer. Another study of nurses over a period of years showed that the ones who were moderate wine drinkers had a 50% less chance of developing diabetes.[12] Wine is better for diabetes and syndrome X immediately as it's not a sugar rich juice. In order to create alcohol the yeast has converted the sugar leaving only the alcohol and health compounds behind. Red wine is thought to be helpful in the protection from insulin resistance and its treatment.

Red wine has been shown to affect eNOS function.[13] Endothelial nitric oxide synthase, or eNOS, is an enzyme that if working prevents insulin resistance and defects in it are believed to cause metabolic syndrome X. Alois Jungbauer from the University of Natural Resources and Applied Life Sciences, in Vienna, found that Austrian red wines are four times more potent than the diabetic drug Rosiglitazone. This drug had global sales of over $1 billion every year. His team focused on PPAR-gamma - a protein that is involved in the development of fat cells, energy storage and the regulation of blood sugar. The same protein that is targeted by the drug. The polyphenols in the wine make the PPAR-gamma more sensitive to insulin. These polyphenols showed themselves to be four times more potent than the diabetic drug Rosiglitazone which also targets PPAR-gamma making it more sensitive to insulin and increasing glucose uptake.[14] This research indicates for all diabetic and metabolic syndrome X sufferers that wine holds some hope for them and their condition. Those who have a higher risk who have not been diagnosed need to be drinking wine especially if a family member has been diagnosed with diabetes or syndrome X. This can reduce their risk of developing this deadly and debilitating disease. Whereas the drug has side effects that cause risk to the heart - wine does not.

Cancer Prevention - Bioflavonoid Punch

Cancer, simply put, is a cell that does not know what it is. One in three people will now die of cancer and with the level of research going on, we have more new cases and twelve types. One of the biggest killers in Scotland is bowel cancer. There are many factors that contribute to cancer. Some are genetically based but some can be induced by your diet. Free radicals and chronic inflammation and stress damage cells. Some say people with depression have a higher risk of developing cancer especially of the breast. These cancer cells can then hide from your immune system, and when they form a cluster become a tumour. Many are full of yeast. An acidic diet can induce the disease and help the cancer proliferate in the body. Much of fast food is acid forming and will also induce weight gain by generating body fat to buffer the acid of the system resulting in obesity. Dr Contreras from the Oasis of Hope Hospital in Mexico who has treated over 45,000 cancer patients said that the greatest cancer is not living your life. Contreras uses both orthodox medicine and alternative medicine to treat his patients.

Wine drinkers have been shown to have less stomach ulcers which are generated from the bacteria H. pylori.[15] Everyone knows that those with stomach ulcers have a higher chance of developing stomach cancer. Wine has also been shown to reduce the risk of oesophageal cancers by 56%.[16] Resveratrol, at just the levels found in a glass of wine, have been shown to destroy oestrogen metabolites known to be connected to forming breast cancer in a study based at The University of Nebraska that exposed human breast cells in vitro to resveratrol.[17] Red wine has been shown to also guard men against prostate cancer, in that men who drank four to seven glasses of red wine per week were only 52% as likely to be ever diagnosed.[18] Resveratrol has also been shown to cause the death of prostate cancer cells.[19]

Cancer may start as an individual tumour but can spread which is called metastasis and generally is fatal. Basically you start to ferment and tumours if you would like to know are full of yeast. Cells from the tumour find their way into the blood stream. If the immune system does not destroy them they can end up in different parts of the body. The cells will find a place on the blood vessel in another body location to lodge into. This could be a damaged part due to oxidative stress of free radicals and ageing. The cancer cell sticks there and starts another tumour. One of the ways to stop the spread of cancer is to reduce the oxidative stress and improve the health of the blood vessel walls. If this is healthy, through the bioflavonoids in wine maintaining the health of the blood vessels the cells won't stick and spread. This is one hypothesis that could apply to the health compounds of wine based on a study in Tibetan Herbal Medicine. A hospital in Israel found that a herbal formula called Padma 28 a Tibetan Medicine herbal panacea stopped the spread of breast cancer on a similar basis.[20]

A mineral found in wine, selenium, has been shown in a number of studies to be a potent weapon against cancer. The University of Arizona found that taking 200mcg of selenium daily resulted in 63% fewer prostate tumours, 58% fewer colorectal cancers, 46% fewer lung malignancies and overall a 36% decrease in cancer deaths.[21] They also found that people who had high levels of selenium in their blood plasma stayed in remission. Seattle's Fred Hutchinson Cancer Centre has seen that a glass of red wine a day will reduce your risk of prostate cancer by 50% and by 60% of the more aggressive forms.[22] Selenium is one of the minerals found in wine.

Digestion - The Acid Kick Start

Your digestive tract is about 5 meters long on average and all along it digestive processes take place. In the stomach acid is created and bile pours in from the gall bladder to digest fats with other enzymes created by the pancreas. These all go to the small intestine where the food and nutrition is absorbed. Then into the large intestine which contains over 90% of all the cells in the human body many of which are lactobacillus or yoghurt bacteria as found in dairy products. These helpful bacteria are also known as probiotics.

Wine traditionally has been used as a sterilising agent for water. Many references to this appear in medical texts and also literature like Shakespeare. Drinking wine safeguarded people from food poisoning. One part red wine was used with three parts water and made the water wine mix safe. Maybe much of the wine that was drunk in the old days was in this form - diluted.[23] The U.S. has over 50 million cases of food poisoning a year. There are even more cases in the developing world. Wine could help avoid this.

A study at University of Missouri in the US showed that Cabernet, Zinfandel and Merlot could inhibit very dangerous bacteria such as E coli, Salmonella and Listeria.[24] Helicobacter pylori the stomach ulcer causing bug was also shown to be inhibited. Very interestingly the researchers saw that the wine only inhibited the growth of pathogenic bacteria and not the probiotic types. Wine also stimulates digestive juices - gastric acids - and therefore appetite through succinic acid.[25] The stimulation of digestive juices also aids the body to digest food. Wine causes the stomach to produce an enzyme that breaks alcohol down which stops *H.pylori*. Dr. Andrew Hart of the University of East Anglia's School of Medicine found that moderate wine drinkers had a 30% reduced risk of developing gallstones. If the stone blocks the duct leading to the digestive tract, as many do, you're talking immediate surgery and the loss of the gallbladder.[26] The bioflavonoids in wine also help the health of the digestive tract by toning the digestive wall.

Many of the anti-wine drinking debates are based on its dangers for the liver. Yes, alcohol in large quantities can damage your liver. Certain parts of the medical community create hysteria about wine, when really their 'liver enemies' are sugar, beer and alcopops. Let's be honest and say that many of the admissions in the accident and emergency departments in the UK, at least every weekend, are from binge drinkers. They have not been drinking wine nor do they offer tasting notes. At the University of California - San Diego they have found that moderate wine consumption not only protects your liver but also guards it against NSAH - non-alcoholic fatty liver disease. The study showed the risk for liver disease was cut by 50%.[27] Over 70% of people with diabetes also have fatty liver which is interesting to note.

Bones & Saint Emilion

As we age our bone mineral density and strength decreases. Bone mineral density in men of ages 50 to 80 who drank wine has been shown to be higher in the 'European Journal of Clinical Nutrition' than those who are teetotalers.[28] When women cross into menopause something very interesting happens with their stomachs. Over 40% of them turn alkaline. This is also relevant to both sexes. As they all age the stomach acidity decreases. Wine helps build acidity in the stomach. For people to absorb calcium from their diets it needs to meet the acid in the stomach and seeing that it's not there this may be a contributing factor to osteoporosis. A French study on wine, women and bone density by EPIDOS in France found that women who drank wine moderately had higher bone density.[29] The study was made up of more than 7000 women. Obviously, the best wines to have are from the soils with minerals too. The acid of wine and the gastric acids it stimulates can aid digestion which will then create the nutrient building blocks of bones that need to be absorbed by the body.

Muscadet, Weight Loss & Diet

You just have to take a quick look at slim French women to see the effects of wine and a normal diet. Dr. Lu Wang from the division of preventive medicine, Brigham and Women's Hospital Boston, saw from a study of women over 39 in age, that moderate wine drinkers had a 30% lower risk of becoming obese.[30] Wang and his team studied 19,000 women over 13 years to see who gained weight and why. They found that women who drank wine moderately did not gain weight compared to the teetotallers who drank mineral water. Scientists believe that it's due to the fact that the calories in a glass of wine get burned in the liver in a different way to the same in cheese which encourages weight gain.

If you think that by drinking wine you can eat a pile of junk food the answer is no. Wine can help the diet but it's best that your diet is more a Mediterranean diet and with fresh food. In addition if you can keep carbohydrates to a minimum especially potatoes, pasta and rice. There is nothing more fattening than potato particularly roast potatoes. Not only are they Bruno's enemies "Ich bin Superstar" they are yours too. As wine stimulates gastric juices there may also be a connection between wine helping us digest our food better and therefore reduce weight gain. One of the other effects of wine is that it may help people to taste their food better. This may also encourage healthier options. We all know how much salt is in processed foods. One company in the UK as part of their healthy food drive took salt mountains out of their food production processes.[31]

Will wine clash with these salt laced and trans fatty acid foods and possibly bring fast food junkies and their palettes back to their senses? If people who have a bad diet full of fast food begin to drink wine moderately, this could also cause them to naturally change their diet for better quality food and a study should be done to see if this hypothesis is correct.

Wine culture is about quality not quantity and perhaps encourages this about food too.

Wine Preventative Medicine Prescriptions

Wine has always been seen as a universal medicine and as a medicine it requires a certain know how, moderation/dosage and type. To include wine as part of your daily health plan for preventative medicine and to boost your wellbeing is a wise decision. First consider which of the four temperaments you are:

Sanguine	Air	Champagne
Nervous	Earth	Medoc
Bilious/bile	Fire	Sauternes
Phlegmatic	Water	Muscadet

The wines in the St Emilion regions are high in magnesium. Champagne is for hypertension as the phosphatic salts will act as heart toner. Some white wines act as brilliant diuretics particularly Sancerre with its alkalising carbonates. For kidney stones you need to take two glasses of the wines of Ripley region of France as they are the most diuretic of all. Or extra dry Champagne that has potassium tartrate. White wines have potassium sorbate which stimulates the production of bile. This breaks the stones down and works as a powerful diuretic for detoxification. Gout is interestingly also higher in areas where wine is not consumed. Arthritis should be treated with Corbieres especially wines from the Minervois region which is high in manganese. This element slows the formation of urates that can increase inflammation.

Beaujolais are high in oenidol which is anti-bacterial.[32] If someone is convalescing it's Bordeaux wines rich in ferrous salts and/or Champagne if there is fatigue. When it comes to heart disease prevention it has to be the Tannat grape wine from the south of France as they have the highest levels of proantho-cyanidins, in particular oligomeric procyanidins (OPCs) according to Professor Corder. Muscadet works very well on cellulite according to renown and Rieslings do very well in helping diabetes. If there is mineral deficiency Saumur wines are your best bet as they are known for their minerality. If you are anaemic then you have to go for port wine which is high in iron and has always been traditionally used in its treatment.

Vine leaves have been used traditionally to treat uterine haemorrhages and the vine has always been used to treat circulation problems in herbalism. Dr Maury thought we should take wine as a supplement quite like how we take vitamins to protect our health and guard against disease. My type of doctor. If only all doctors could also include wine in your treatment? As a result of this book let's hope they will.

Champagne & Alzheimer's Prevention

Well I started the book on the fact that I do prescribe Champagne for depression. So it's only proper that I explain why Champagne has health benefits. Dr Tran Ky, a Professor of Urology in France and an aristocrat of Cambodian extraction, has written 7 books concerning wine and health. One of his most recent was about the health benefits of Champagne called 'The Healing Power of Champagne.' In it Dr Tran, who apart from being an excellent physician is also a talented science communicator explains the connection between Champagne and health. Champagne, as it turns out, is quite a health panacea in moderation - as in a glass a day. The Sun King it appears should have stuck to bubbles.

One recent scientific revelation is that Champagne has a great effect on the brain. Champagne production uses a mix of both red and white grapes. Parkinson's disease, Alzheimer's and stroke are all prevented by Champagne. Blood flow is improved from Champagne.[33] Scientists from The University of Reading examined the brains of mice - one group fed Champagne extract and one without - after the effect of stroke.[34] They found that the brains exposed to Champagne, particularly the neurons, had been given protection by the polyphenols. Scientists believe that the compounds may also give protection against Alzheimer's and Parkinson's disease.

This is very exciting news in that so far the health benefits of wine were just connected to the heart, vascular system and red wine. Champagne wine has now been connected to the brain and also with an anti-inflammatory effect and toxin expulsion from the body. Dr Jeremy Spencer saw that two polyphenols found in red wine, specifically caffeic acid and tyrosol, have antioxidant and anti-inflammatory properties. In addition they also cause the body to excrete toxic chemcials.[35] The scientists also found that the compounds guarded against neurotoxicity and even crossed the blood brain barrier.

For centuries Champagne was prescribed as a tonic and there is scientific evidence to also show this to be true. A research group at the Laboratoire de Biochimie in France looked into the level of neurotransmitter stimulation between Champagne and still wines.[36] Six to eight minutes after drinking a glass of Champagne the wine causes a discharge of mood enhancing neurotransmitters like dopamine and serotonin occurs. This may explain why French psychiatric hospitals gave depressives a glass of Champagne a day to boost their spirits. And why I also follow the same tradition.

The same group went on to show that Champagne drinkers really were also shown to have higher levels of nitric oxide that controls blood pressure. Therefore they will also benefit from a reduced risk of stroke. This is yet another example of just how powerful the role of wine, as a remedy, is in health and wellbeing even with Champagne. I wonder what health benefits Prosecco holds for us? Sparkling wines are a huge growth market, and all those drinking it will be happy to know that they are protecting their brains and heart as they do.

The Higher Health Potency of Whites

Most of the health benefits of wine are normally attributed to red wines as many of the health compounds in wine - polyphenols - are found in the red skin of grapes. But white wine also has many health benefits. Now red wine may contain higher levels of phenols but studies have shown that they are actually more active in white wines in inhibiting bad cholesterol LDLs.[37]

In another comparative study, white wine was better at reducing blood pressure than red wine.[38]

The Jordan Heart Research Foundation found that free radicals were reduced by 15% in red wine drinkers and 34% by white wine drinkers while red wine drinkers experienced a reduction in the blood's clotting ability of 10% and white wine drinkers 20%.[39]

In a letter to the Editor of the 'International Journal of Food Science and Technology' titled 'Free Radical scavenging abilities of beverages' Troup et al pointed out that "if the health promoting properties of wines are related to their superoxide-scavenging abilities, then white wine is at least as effective as red". He saw that the antioxidants in white wine were smaller than those in red in terms of molecular weight. Seeing as they are smaller they are able to get to more places in the body which contributes to their effectiveness.[40] Microbiologist Mark Daeschel from Oregon University has found a novel use for white wine - wine cleaner. He is proposing to use waste white wine as a household cleaner. He says that as it kills salmonella in a matter of seconds and that it's a very viable and useful kitchen product.[41]

Sun protection creams from wine? Well it isn't a million miles away. Scientists in the University of Barcelona found that compounds in wine protect the skin from damage from UVA and UVB radiation.[42] Mainly that the compounds protect from skin cell death and damage from exposure to the sun. White wine also works as a great diuretic due to these potassiums it contains and thereby can help in the reduction of high blood pressure.

Wine & the Relaxation Response

Many people say that one of the main reasons they drink wine is for relaxation. Wine they say helps them to relax and unwind from the stress of the day. We now know that over 60% of visits to primary physicians are for stress related problems. As much of 75% of all diseases are inflammatory diseases triggered and made all the worse from stress. Asthma, eczema, ulcers are all inflammatory diseases. The body has two biochemical states - one of stress and the other of relaxation. We are all more familiar with the stress 'fight or flight' system than relaxation which as you can see causes many diseases. If we perceive danger we immediately trigger the fight or flight response which causes us to release from our adrenal glands, just above the kidneys, the stress hormones of adrenalin and cortisol.

Adrenalin increases your blood pressure and heart rate whereas cortisol interferes with your immune system, increases blood sugars, stops digestive processes and heightens a feeling of fear in the brain. The vagus nerve – a 200 neuron cable trip switch - beginning in the brain and reaching down to the stomach is deactivated. Your mouth goes dry as your wiring is built to run from danger. Running from a lion you may not have the time to swallow saliva. A stressful situation will not do as much damage to the body as a stressed person who is constantly stressed. Who perceives the world as stressful? There is no such thing as reality only perception.

We now know the level of diseases that stress biochemistry induces: hypertension, PMS, insomnia, cancer, heart disease and anxiety to name a few. Stress also causes us to comfort eat and in particular, fast food and high calorific diets which cause further harm. The natural solution is to learn how to induce the opposite response of the body biochemistry which naturally restores and rejuvenates your body. Large quantities of stress especially when combined with hypervigilance over a period of time will kill us.

In 1975 Dr Herbert Benson, a Harvard cardiologist, published his book 'The Relaxation Response' and his solution to the stressed state. This bestseller opened a new understanding of the mind's influence in health which still grows to this day. Recent research has also shown adults who practice the relaxation response have thicker brains as they age. This is innately hard wired in the body and he described it as, "a physical state of deep rest that changes the physical and emotional response." This state, also known as 'rest and digest', activates the region in the brain called the hypothalamus and releases neurochemicals that counter the fight or flight biochemistry. He developed a way to teach people how to induce these states from studying people who meditated. Prayer, yoga and breathing deeply can induce the 'rest and digest' biochemical state, the benefits of which are reduced blood pressure and anxiety. Wine can enhance this relaxation response and should be studied to see to what degree. More studies should be done on how wine reduces stress and induces the relaxation response and these should be both psychological as well as medical.

If a glass or two of wine a day are found to help induce relaxation this could benefit medicine as a whole. Wine could aid in reducing and preventing diseases that the 'relaxation response' successfully treated such as:

- High blood pressure
- Anxiety and depression
- Infertility
- Insomnia
- Menopausal hot flashes
- Many pains - backaches and headaches
- Phobias

A glass of wine could induce a relaxation response by being a mental anchor, as they say in

NLP, to mark the end of the working day. This could cause the biochemistry to switch to 'relax'. In Japan, one wine brand has been made with a neutraceutical to increase the relaxation response. The market research studies showed that people drink wine to relax and the Japanese listened. The compound in question is added by Mercian Wines Japan to white wines and is called GABA (gamma aminobutyric acid) which has mood improving qualities inducing relaxation and balancing the brain. GABA also works by stimulating higher levels of human growth hormone that will keep you younger, in that it will help to reduce fat uptake and boost muscle quality. The wine has 8 times the levels of GABA of normal white wine.

If over 55% of people, according to some market research reports, are saying that they drink wine for relaxation then we need to study why that is. We need to understand this scientifically to make perhaps an even greater case for wine and health. If one glass of wine can trigger a response in the body biochemistry to interrupt the production of stress hormones that will have huge implications for public health and medicine. Moderate wine drinking could be seen to be assisting people to avoid stress related diseases. Moderate wine drinking might get them to give themselves more time for relaxing. They may even take time to relax. The popping of a cork, the colour, the aroma, the taste and the aftertaste, our whole being's appreciation of wine must make our innate relaxation response drool like Pavlov's dog. We just need to prove it.

The Future of Wine & Health

Well we have seen that one compound in wine synthetically made with a 'tweak' is now worth $725 million. This is one of 1,000 compounds naturally occurring in wine. Even more in nature. Many research groups are investigating resveratrol for its health benefits. But what about all the rest? The population studies are extremely helpful in making the case but what about more biochemical studies? In an age of metabolic syndrome wine may also be a healthier option if it stops people drinking fizzy sugar drinks.

The wine industry and groups need to do more to encourage people suffering from heart disease, diabetes and cancer to start drinking wine to help their health. This should possibly be part of a national health programme which I would like to see happening especially in Europe due to all the health benefits. This could save EU member states millions and reduce hospital waiting lists. Women who hit the menopause and who start to take supplements should be encouraged to drink wine for their bone health especially for its ability to increase bone mineral density. We really need to get more and more doctors prescribing wine.

Many hospitals in the USA do serve wine but how about serving wines for the particular health compounds that we know affect particular diseases. How about medical sommeliers? In light of the culture of moderation that wine has why don't governments start wine appreciation evening courses at universities? Many of the beer companies marketing people already run promotions there. In secondary schools we need to have older students study the health benefits of wine as part of their biology and health classes in national curriculums. This will encourage young people to have wine as their first alcoholic drink and as a result drink moderately. This perhaps could be a great social benefit and may also help reduce, within a decade, the stress on emergency departments and the demands on hepatologists who have to deal with liver damage from beer and spirits. Not to mention the security risk and rates of assaults on medical staff from beer and cider intoxicated patients.

The wine industry really lags behind its siblings, beer and spirits, in terms of innovation. This needs to change in terms of wine and health as the innovative steps could have huge impacts on public health without affecting taste. Functional foods are a big trend and help to society. This market is worth $176 billion with added vitamins, minerals, herbs and other natural health compounds. The 'functional food' wines may not end up being from a Chateau but could have a big impact on people's wellbeing.

So there is no excuse anymore for the industry to stall in this area. Innovation, adding functional foods or neutraceuticals like Mercian, for the consumers and to boost the innate health benefits of wine drinking is a must. Excuse the pun.

Two glasses a day is what is meant in terms of wine and moderation. Yes, anything after starts to create problems but those who stick to this would be doing a great service to themselves. At present many people in the UK only drink wine once a month. There is lots of room for more, especially in light of all the scientific evidence above. With more to come as more studies explore one of the world's greatest natural medicines gifted to us by the Greek God Dionysus.

6

Artificial Intelligence Vineyard

Somewhere up there in the sky, nearly 900 km away, a satellite belonging to the Oenoview Project called Formasat 2 watches over vineyards. The satellite which is owned by Taiwan but built by Astrium in Europe is always in orbit policing the environment.

If you saw the movie Artificial Intelligence that may have been your entrance into this realm of thinking machines. For me, it was over a decade ago when I was working as a biotechnologist for The Body Shop International on a number of technical projects for the founder Dame Anita Roddick. Near the waste water treatment facility at the back of the water treatment plant was a greenhouse. When I went to investigate, I walked into the world of aquatic ecosystems, living machines and living water. These 'living machines' are used to treat waste water. They return the water through an intensified natural process of plants - especially the food web in the roots and full aquatic ecosystems - to water purer than your tap water.

Here you could see nature was alive and could balance itself. I started to see how much and what type of effluent it could handle: shower gels, face creams and toners. A real complex chemical mix actually. One could see that the ecosystem was able to deal with it all and the pilot project was developed further to handle the total amount of waste from the factory and ended up on Discovery Channel. The company then installed an aquatic living machine at their headquarters to handle most of the waste water. This way they return to nature water as good as they received it.[1] I wish more companies treated the environment this way, especially oil & food companies, breweries and wineries who regularly dump into rivers.

One of my ideas was to connect the 'living machine' to a computer system that would have been self-regulating and for example, if it was hot the windows would automatically open. You could say I stared to see it as an artificially intelligent living machine. I did not know it at the time but I was moving into the realm of AI. Artificial intelligence is and could be used by vineyards to regulate the whole ecosystem that is taking place in the vineyards. When it's dry, water could be sprayed. The sap could be monitored. Today with apps and smart phones the wine maker can look at the vineyard by square inch and remotely. An ability to zoom in and zoom out that was hitherto only a faculty of eagles flying in the sky.

Uber Precision Wine Making

With the changes in the environment, the wine industry is becoming more high tech and with recent developments in mapping, vineyards are employing this technology increasingly to ensure a fine harvest of the best quality. In California, for example, over 400 vineyards are recording weather data.

One of the pioneers of the high tech vineyard is Joy Sterling a second generation wine maker who grew up in Paris, went to Yale and became a TV news reporter for a whole range of interesting assignments. The Sterling Family are one of the leading wine making families in Sonoma County. A striking blonde lady and brainy to boot who has travelled the world with a natural attraction for adventure. During their time in Paris the family had a roller coaster tour of culture and went out there and engaged with Europe, led by the parents. This is probably the phenotypical cause as to why she has jumped into what she calls precision farming. Joy is the generation of wine makers that come to the trade with a whole host of extra skills from other industries, in her case journalism, that benefits and adds value to the industry and their wines. Plus her adventurous spirit for new technology.

Precision to Joy means every square inch of her vineyard being optimised. The cost is mighty but Joy is very clear as to the reason why. "The only thing that is proprietary to us are our grapes. All of the wine making technique in the world is universal. So the only thing that we have that we can claim as our own are the grapes and whatever unique flavours we can pull out of the ground. That is what precision farming is about, precision farming has to do with exactitude." She knows this well and as a result her vineyard has repeatedly won awards. Some say this sparkler is one of the best there is. Therefore you need to optimise this to stand out in the crowd.

Iron Horse winery is mainly known for sparkling wine and has even been served at The White House, most notably at a function for peace attended by Mr. Gorbachev. As a trustee on The Leakey Foundation, she has a deep understanding of the issues affecting the planet today. Their sparkling wine has been served for the past five consecutive presidential administrations.

Joy and her team at Iron Horse using this technology know every vine on her estate. Using technologies such as GPS mapping, GIS (Geographic Information Systems) and infra-red satellite readings they get vines that produce grapes that ripen evenly. Joy tells us, "All our viticultural practices, all the pruning, canopy management, irrigation, cover crop and even harvesting decisions are determined on a practically block-by-block basis." If you go visit them you will see the wine makers walking around the vineyard with baseball caps and PDAs connected to GPS readings of the vine locations they are working on. By nailing down the data and mathematically modelling all the variables, their chief wine maker David said," We get exactly the grapes the wine maker wants. When you're making one of the world's finest sparkling wines that is in high demand each grape really counts."

Wine from Space - The Oenoview Satellite

This is the first remote sensing high resolution satellite with daily revisits across the earth. The satellite is an important environmental one, as it is used to map agricultural areas right across the globe and its oceans. Even disasters like the Tsunami and oil spills are monitored using its technology. The daily high resolution and extremely detailed images both in black and white and in colour, that the satellite generates are easy to obtain which is why Oenoview was started.[2] This French based project, in connection with the European Space agency, generates infrared satellite images to help the winemaker. The technology recently won an award at the Paris Agriculture Show. When you look at the images of vineyards from this data you can see a rainbow of colours in an area of vines that has not grown uniformly. So you see areas of green, yellow, orange, red and blue. If you see brown the vines are under stress. The famous Robert Mondavi Winery successfully managed to stop a disease using this type of technology to counter an attack from phylloxera aphids. They saw the change in the infrared images from NASA and immediately jumped into action.[3]

Anything that generates heat emits infrared radiation and this is what the satellite captures using its sensors to generate images. The images produced allow a technologist to quantify per square meter of vineyard: the vigour of the vine, the amount of leaves per square meter, grape bunch size, grape weight, soil water levels and minerals, all at different parts of the field. This, as you can see, gives these wine makers tremendous leverage in their practice of viticulture and management of vines. This is like how the Bristol police catch illegal cannabis growers using helicopters. They just study the hi-tech heat reading from the houses in residential areas. The ones with the very hot roofs are harbouring cannabis plantations.[4]

The colour images of the vine zones, technologists now know from their studies, give a particular grape and taste. The image also allows vine vigour to be mapped from zones of strong and weak leaf growth. This knowledge of vigour per m^2 allows estimates of yields to be known. If the images start to show the grapes changing colour the wine maker can start to remove leaves for the sun to mature the grapes. There are just so many possible uses with the information.

Probably the most important benefit of this technology is at harvest. The picking can be organised from the image. Traditionally harvesting of grapes is done in one session. They are then separated into batches as some will not be of a quality for wine production as they are too young. Now with hard data from these satellite images, the grapes are harvested in seven sessions with four days in between, to let the grapes mature. This practice optimises grape quality for wine production.

One of the main wine scientists working on the Oenoview is also a world renowned expert on the Grape Berry Sensory Analysis tool which is now an industry standard. In fact he is the inventor of the system. He devised a system for 20 characteristics of the maturity of grapes in a common language. Over 5,000 vineyard specialists have been trained in the system since it was launched over 15 years ago. Who better then to tell you when to pick and when not to pick your grapes than innovator Jacques Rousseau?[5]

For wine co-operatives spanning large areas of viticultural land, this is extremely useful in both waste reduction and optimising wine quality. Ideally an image will be taken just before summer and at the start of the vine growth cycle. This is specifically assessed for the implementation of pruning, watering etc., then in August the image will show in which way the grapes should be harvested or how to harvest the mature grapes first. Scientists started with an experimental survey of two vineyards to calibrate the system. Using the daily images the scientists can detect change even over large areas. All this data allows the wine maker to access in matter of hours what would normally take 30 years to do in an empirical way. Impressive? This also means you can pick the right bunch at the right time. You can send your picker out with the exact co-ordinates of where the bunch is.

Due to the daily visits, scientists can monitor areas of land frequently and build up data quickly. Using these aerial images and satellite pictures, scientists are able to decide which grapes are ripe for picking, for those who are lucky enough to be connected to the Oenoview project. The infra-red satellite readings give very precise data on temperatures. The technology can tell them what the levels of water in the soil are. Add to this the leaf surface area and bunches of fruit, you begin to get a better understanding of where and why technological winemakers like Joy are going. These systems can also alert them to any disease right across their growing area. Forewarned is forearmed.

The Fruit of Fruition Sciences

Take a bunch of lads who are French oenologists, plonk them in Nappa Valley California, throw in some sensors and you get Fruition Sciences - a new and exciting viticulture technology company. The two founders Thibaut Scholasch and Sébastien Payen trained in France before moving to Nappa, where a number of high end vineyards are utilising their technology. They won the Imagine 'H2O Water Prize' for the kit which has been set up to encourage better use and management of water through innovation and entrepreneurship. Their HQ is in Montpelier France but they operate in the USA and they are now moving into the European market.

According to Thibaut, animals have a heart to pump blood around the body and the heart in vines is the interaction between the sun and the vine causing transpiration - plant water circulation. Like the heart in humans under exercise, if you run faster the heart beat will be faster and it's the same in vines. If the plant feels its water, or what some call sap, is limiting at the roots, it will modify its consumption. By monitoring the plants water use you can maintain its health, manage your water use and improve the fruit quality. The plant has a way of managing different climate conditions in order to produce fruit, such as high winds and heat spells that are not part of extreme climate. Without knowing this information precisely there is no way of knowing what's going on exactly in the vines.

Without this the guys at Fruition Sciences tell me that wine makers will over water their vines. In an area like California where water is in shortage and over 80% is used in agriculture, this is not sustainable. The industry needs the hard data that their system can provide.

The exceedingly clever guys at Fruition Sciences have come up with a handy piece of kit that can enable wine makers manage water in the vineyard. They have been on the national news in France and in a number of publications. They add a heat sensor to the branch of the vine which has a thermocouple that measures the flow of water by the plant. This small sensory unit measures about 3 cm along the branch. The sensor is wrapped around the branch and picks up the temperature at each end. As the sap moves up the plant it displaces heat. This temperature difference is recorded and from it Fruition Sciences can tell you how much water the vine is using and how much is flowing. A slow rate will affect the vine detrimentally and most importantly the fruit. A fast rate dilutes the grapes and their quality in terms of colour and taste. So Fruition Sciences helps the viticulturists to control what they call water deficit. On average they place four of these sensors in a vineyard.[6]

The team tries to help the vineyard to manage this deficit exactly in order to reduce water use and encourage the vines to grow optimally for wine which means smaller fruit but better colour and taste in the form of tannins. The data from Fruition Science can be used to mathematically optimise the industry and its water needs. This will enable us to understand vineyards scientifically better and how the vines work in terms of weather and climate with which we can improve the world of wine. The technology takes some of the grey areas out of viticulture where we had no real time data. This temperature difference is relayed back to Fruition Science's data centre via a data logger located at the vineyard. This weather station also possesses a solar panel and radio wave emitter that sends data back via the Internet. The solar panel produces the heat of the sensor as well as power for the data logger and radio transmitter.

The device logs data from four locations in the vineyard for analysis. The data loggers are connected via insulated cable to the sensors on the vines. The data is computed by Fruition Sciences and posted on their website for the vineyard managers to see through an individual access code. The data is from continual monitoring twenty four seven. From this information the vineyards can water, or more importantly not water, the vines depending on the vines' needs. This form of vine management aids the winery in producing smaller grapes on the vines of better colour and higher tannin content. By reducing the amount of water a plant uses you can influence the production of fruit.

By withholding water the vine will produce smaller fruits which produces better tasting wine. In terms of climate change winemakers will need to consider what is happening in the environment in growing grapes.

All of these systems, Oenoview, Fruition Sciences and data from the weather stations now joined with the rate of sap use, brings us closer to the possible creation of the AI vineyard. Temperature and water levels, humidity are all sent back to a central computer for the viticulturist to monitor. This information is highly valuable to know when to harvest and to influence how we grow and harvest wine in the climate challenging future years.

All these people are now collecting a whole host of data that we previously did not have. This data can be used to help in the production of wine for the end user without interfering with nature but working alongside it and in harmony with the inevitable environmental changes. The vineyard may become slightly self-regulating. For example this data could be fed back to a computer that will water the vines automatically if the sap rate is falling dangerously low. All of this data fed into a central system can, as we get closer to seeing the AI vineyard, produce fruit optimally with the use of modern technology that dances with the forces of nature and adapts to specific climate changes to safeguard the world of wine.

Grape Picking Cybernetic R2-D2s

Everybody can remember R2-D2's debut on the cinema screen in the movie Star Wars. The idea of an R2-D2 inspired the mind of the boy I was. Where was R2-D2's mind in the machine? The possibility that one day there may be a robot like this with some form of intelligence and consciousness captured my imagination and never went away. With all of the data that we now know about vineyards, the vine and the viticultural practices of when and how to prune, we could benefit from a robot. Couldn't we? Vineyards can be very labour intensive. There is no reason not to believe that in a couple of years R2-D2-like robots will be running the vineyards' daily needs. You would have read earlier that robots have started to even taste wine. San Diego based robot technology company Vision Robotics have developed a vine pruning and grape picking robotic arm.[7] A bit like a vineological butler for the vineyard. For fruit picking robots, strawberries interestingly are the most difficult for them to pick which goes to show that grapes are not the only fruit on the robotic picking agenda.[8]

The mathematical data that will be gathered by these robots could also be plugged in to an intelligent system. There would be 24hr carers and even grape pickers for the vines. What better care for the vine could we give? The robots could be remotely connected to information sources such as GPS, satellite images and water levels. Using this information they could set to work in precision viticulture to produce the best grapes for the greatest of wines.

Who's to say that in the future we might not find planets that are ideal for wine growing? We will have to wait to see what the latest probes find on Mars and 20 years more to see what the first space mission to Mars brings back.[9] Our ET or extra-terrestrial friends seem somewhat keen on the idea. They left an interesting 'crop circle' design at Malvino Family Winery to say so.[10]

Then we will possibly be living in the age of planetary terroir but that is another story.

7

SEO Wine 2.0

Each year there are 14 million conversations taking place on the Internet about wine according to Vintank.[1]

In one of the hours that you are taking to read this book, over 1,600 conversations on wine will have taken place in social medias such as Facebook, Twitter and YouTube. Despite this over 96% of vineyards in California have not yet jumped on the wave of these opportunities to engage with their customers.[2] Neither it seems have the French and many other producers worldwide.

This vinocentricity has to change and will on the back of who the wine enthusiasts are and what they will expect in the future. If you don't think that social media has an effect on wine sales then talk to the Cup Cakes Chardonnay brand people who have been riding the trend. The success of their brand has been driven by social media, a trending keyword and a market for the wine Cupcake consumers want - Chardonnay. Until the arrival of the cupcakes trend the winery did not exist. The arrival of Web 2.0

social media is really what the wine industry is going to have to rapidly come to terms with in relation to customer relations management at a high standard.

Web 2.0 is a lot more than just blogging and more a whole collection of social media websites whereby communities of people gather together. Using the media and especially subscribing to different types of media special pages such as in YouTube, Twitter and Digg, information can spread in a viral way with the touch of a button. Let's say you have a blog with 3,000 subscribers. Every time you publish a new article you can notify all of them simultaneously. They may have social media of their own such as Twitter and Facebook. With one touch of a keypad on a personal computer and even using their phone all their friends or wine loving community can also be informed and so on. Within minutes your article can be notified to thousands of people and posted and replicated on thousands of relevant sites. A fermenting exponential information dissemination. Such is the power social media yields. Whereas in the past the public was excluded they are now a core part of what media is and does. A new phenomenon called 'flash crowds' is as a result of Twitter whereby people are invited to gather via twitter messages and within hours a crowd has formed. This happened in London when Michael Jackson died - thousands of people suddenly gathered together to sing his songs. A more sinister side is when flash crowds cause a riot.

You may not need to be told how many people are using the Internet but you may need to know that they are using it for more than info searches and porn. The revolution of Web 2.0 is a phenomenon that the wine industry can grow with. Especially when so many Millennials are 'digital natives' who use the net to connect in a way that previous generations used the phone for. The one attached to the wall. The web started with let's say static websites and email. Don't forget if Facebook was a nation it would be the fifth largest in the world. Arbour Mist - a fruit wine - recently amassed, with some clever social media marketing, over 270,000 fans in a very short space of time.[3] The label life cycle was renewed.

The dawn of Web 2.0 caused a revolution in terms of what users could do with the web, or better still, what the web could do for them. In the past the media model was the newspapers creating stories that fed the public. Now its news editors monitoring blogs that we create which then end up in the world press, sometimes even on the front pages. If the information is wrong it can do a lot more damage than the old broadsheets, or if it's right as in the case of a senator, it can do damage much faster with a few tweets. Spreading far and wide. More and more newspapers have shut down as a result of this blog and twitter movement. Additionally there are podcasts which can be downloaded regularly to your phone or computer for you to listen to in the same way as a radio show.

Snooth.com has a US-centric, affluent, well-educated audience of 500,000 wine enthusiasts and users who typically visit the site weekly or more. Their research found that:[4]

- 96.9% of survey respondents use Facebook
- 25.1% use Twitter
- 30.4% of respondents visit Snooth on a daily basis
- 81.7% of respondents visit Snooth once a week or more
- 55.1% of respondents are casual enthusiasts
- 33.3% are more committed wine lovers

Pierrick Bouquet one of the founders of My Social Winery is at the interface between wineries and their consumers. His company will engage with those having the conversation and let them know about the existence of your wine. They will also create the 'feed' that is required. The ever important content perhaps even stories about what is happening at your winery. With a name like Bouquet in this industry he has a head start.

"You see it's about people wanting the relationship with the producer that their grandparents had with the local baker. Only now they want it with the bigger brands. The social media conversation about wine is also growing by 10% a month." Pierre did some research of his own and he found from his poll of wineries that 87% don't even have a blog and 75% don't even have a Twitter page.[5] More astounding when you consider these media platforms are free to set up in comparison to big marketing and public relations spends. Free to directly engage with your wine drinkers. One free service Google offers is an email alert you can arrange to automatically send you an email if your winery ends up in the news or a blog. This is very easy to do and will keep you informed about what journalists and bloggers are saying about your wines.

The digital age may also be taking away the embarrassment of asking questions about wine if you are neophyte to wine culture. Due to the distance between people on the internet, as in geographical distance, people find it easier to ask questions. Tom from Seattle can ask Mary from Miami questions without losing face. In the old days of advertising there was a saying that 50% of the spend worked but they did not know which 50% it was. Now we have exact data. You can know who your consumer is exactly. Where they live and their age. Their whole profile. You can do promos to encourage them to get to know and enjoy your wine. Most exciting of all you can get to mail order them directly. This is very feasible for wine co-operatives from French regions to start doing in the USA.

Get in the conversation: this is what the marketeers are now saying. We have come a long way since the pamphlets written by Doctors sponsored to do theses on wine and health in the 16th century. The activities of the Champagne and Burgundy regions form part of marketing history. Communities like wine enthusiasts are now all talking and if Web 2.0 is about anything it's conversation and getting in on that conversation and in our case that is all about wine. Let's not forget wine is very much a social product.

If a wine enthusiast or producer needs to know anything about the internet it's that it's not a thing - it's a place. As busy as Oxford Street or Times Square with a massive flow of people.

To All the Vayneaks Out There

If you don't know what a 'Vayneak' is yet, then you have yet to meet one of the most exciting entrepreneurs the international wine business has seen in many years. He has created one of the biggest conversations about wine on the web. Gary Vaynerchuck and his Wine Library TV are causing major transformations right across the industry. All because Gary - Hey Mom - has gone and done the most obvious, perhaps simplest thing imaginable -

talked directly with consumers. To do this as effectively as he has he harnessed the power of social media, mainly with YouTube and magnetised people and communities from all strata of the web. The wine industry had spent years ignoring them. In so doing Vaynerchuck grew the New Jersey family business from $5 million to $50 in a few years. He has ended up with a best-selling book called Crush It, a seven figure book deal and is represented by the same people who manage celebrities such as Oprah. Vaynerchuck says that today everything is under the spotlight as a result of social media focus. The sad thing is most wineries have yet to implement any social media despite all of the evidence to do so.

One of the most important things to know about the internet is that over half the traffic is seeking video. They prefer video visuals and presenter anchors to explain things. With over 75% of people learning best kinesthetically you can see why Gary was such a success in an information vacuum left so wide by the wine industry. He has even said publicly that there is room for more wine TV on platforms such as YouTube. Gary tapes, produces and shoots the show with a video camera, the cost of a couple of hundred bucks and the help of his family. He now has a huge following in the wine world. Some wine experts call him 'the people's choice' in terms of wine recommendations. He has done so much for wine awareness. In fact he has created over 800 videos on wine, some with the wine makers themselves in the last few years. Vaynerchuck is one of the people who just gets it. He saw the bandwagon before it came and like all true entrepreneurs brought it into being for the better of the wine world.

Wine Blogging Experiences

Up to the arrival of blogging you only had the likes of wine connoisseurs who were published book authors. They were the ones who led the way in wine. Added to this were the wine columnists in newspapers but now you have a whole host of people blogging about wine. A blog is a website that has content, mostly articles, updated regularly. Anyone can set one up. If you build a reputation, you get visits from interested people. They leave comments about what you write on the interactive feature of the blog. They also may tweet it to their friends and notify their Facebook friends. There are two very well attended Wine Bloggers Conferences, one in Europe and the other in the USA, to bring all these bloggers together in one room and to discuss wine blogging, taste wines and visit vineyards. A recent conference in Portugal enabled the bloggers to have a tasting of over 50 grape varietals. Some of these attendees had never set foot in Europe. Awesome. Over 300 bloggers attend each conference and there are well over one thousand good wine blogs to choose from. The 'wine critic' is no longer three people.

One of the most important parts of wine blogging for enthusiasts is the ability to get independent recommendations. This, to most people, is very much the meaning of wine blogs and why they visit them. There are now many people regularly blogging about wine. The major reason they blog is to review the wines, apart from that there is great variety in the writing styles and interests of the bloggers. They are all involved in the same thing - conversation about wine. Catavino a blog based on Iberian Wine has over 20,000 visitors per month.

The Wine Anorak is probably the best place to go for the technology of wine. Dr Vino won the American Wine Blog Awards. People want independent information about wine that is not been spun by public relations consultants. Some people specialise in Portuguese wines whilst other people like to write all about what is happening in the area of organic wine. Some bloggers can earn up to $30,000 a year and anything up to 40,000 visitors a month. Readers can post comments in relation to how they feel about what was written - whether it impressed them or not. In addition they can add valuable information to the discussion. There is a lot more than tasting notes being bantered around. Blogs about new wine entrepreneurs, eco wines, technology, winemaking and biodynamic wine are all available. The best blogs are therefore the ones with the best content which will attract more repeat readers and followers.

Wine blogs is one way that less well-known brands can make it to the enthusiasts without the use of a massive marketing spend. If the wine has great taste, is well produced and has an interesting story to go with it, you could make it directly to the people who would drink that wine. Why? Well because you can see exactly who they are. They are openly subscribed to various blogs and you could approach them directly. Lots of organic wine lovers will be subscribed to various social media sites and if they are, they will want to know about something new in their area. If that blog is high on the search results for 'Pinot wine' your wine should perhaps be on that blog. One business maxim states that you should never expect people to pave a way to your door. This is a proactive effort as Gary has shown and the results of which are truly rewarding especially for small producers to make it to market.

A One Man Show

This is a great example of web 2.0 and its consumer power with over 100,000 wine enthusiasts. CellarTracker CEO Eric LeVine says in his own words "Since its launch in 2004, registration has grown to 126,000 members. On a typical day, you are tracking (adding or removing) close to

20,000 bottles for a total of 21.6 million bottles. The database includes 980,000 wines from 70,000 producers, one of the largest in the world. The CellarTracker community has also emerged as an abundant source of wine reviews with more than 1,600 wines reviewed in a typical day for a total of 1,663,000 wine reviews, all written by real wine enthusiasts. The site is one of the most heavily visited wine websites in the world with 25 million page views per month from several hundred thousand unique visitors."[6]

Eric started with the idea when he was working as a developer with Microsoft. He believed that a database would be useful in the world of wine and to share just with his friends. They all posted the wines they drank and what they thought about them and next thing the system was expanded and just took off. Until recently he has been a one man operation. Another example of an entrepreneur bringing his skills to the wine world to create a new phenomenon.

Crushpad - Pirates of the Garagiste

Ahh haarr. I mean, I mean ahem ahem. There is something rebellious about making your own fine wine and succeeding in doing so. The most exciting venture that has come out of wine people power Web 2.0 is the wine making company called Crushpad. As a result of social media revealing just how many people not only drank wine but who were passionate about winemaking, Crushpad was conceived to give them the ability and facility to make their own high end wines. Hands on, from the type of grapes to the blend and the designer label for packaging. To date they have had over 5,000 clients and this also goes for new wine labels, some of which have received over 90 points from reviewers. For those who also had a dream to make wine or even get into the wine making process without having the capital to buy a European or Californian vineyard, Crushpad is your stairway to the celestial realms of fine winemaking.

The venture started in a garage in San Francisco with a couple of friends who were making quality wines. Taking it to the next step was obvious as the wine enthusiast world became more defined. The USA has a long tradition of making wine in garages as a result of prohibition. The *garagiste* wine was not great, not great quality but when you connect this company and technical expertise through the social media it has developed and the wine making community its nurturing, lots of exciting things are possible. In organic farming over 30% of the new farms springing up have been founded and run by people with no background in farming. Experienced marketeers and accountants from the city are moving out to the country. They all describe themselves as 'recovering' on twitter. This, some commentators say, is the reason why organic food has succeeded so well. Crushpad could be a haven for all these 'recovering' marketeers and executives and if they found their way into this industry and took some of the market share from the big brands, this could be beneficial for the industry as a whole. Not that I have an entrepreneurial bias or anything. Wink.

Using social media, Crushpad is tapping into a large community of aspirational wine makers. By nurturing them with education, technical consultancy and information sharing, they are raising the standard of the millennial *garagiste* wines from what could have been. With products they have developed locked into their personal account on the winery's website, who knows what could happen? The market for those services in that type of winemaking zeitgeist is huge. This could cause so many to become self-producers. Many of them have fathers and grandfathers who did produce wine in the garage. Making wine is in their blood. Can we encourage them back in again with the right tools, technical help and fine ingredients? I bet we can. That would be a real *coup d'état* to see 90 point wines being made in the garage.

Enthusiasts who enter the cellar doors at Crushpad come as individuals or groups. First off, the type and style of wine is deduced through a lengthy questionnaire to suit the person or group. Then a wine plan of 30 steps is produced which includes the grapes, harvest time, processing dates, fermentation, type of barrel, and bottling and label design. Generally, clients make a barrel which is equivalent to 25 cases or 300 bottles. This barrel can cost anything from $5,000 to $10,000 max. Each bottle works out at roughly $25 but if you're producing a 90 point scoring wine, and as good as the great estates, then you're saving money whilst enjoying all the quality and taste. The hard work of making it too. How satisfying.

When the company was seeking capital to expand, rather than go the venture capital route, they approached their high net worth clients. They came up with the $9 million that was needed. In fact this sale of shares was over prescribed. So they have now gone from the garage to a state of the art facility in California to having a European facility in Bordeaux to opening an office in Japan. Imagine what will happen when they land in China? Michael Brill is the Jack Sparrow of this world. Had he said to people a couple of years previously that this venture would not only grow out of their garage but have a high tech facility upgrade and have a similar one in the south of France, they would have said 'Shiver me timbers'. Instead he has discovered treasure for so many wine lovers.

Just look at what the *Garagistes* movement did to the face of the Bordeaux wine in the early 1990s. My hunch, and it's a strong one, is that it's not going to be long before baby boomers start making wine in the garage again. Most of the equipment is in their garages and basements just gathering dust. If at some point Millennials step on the learning curve, they will be teaching each other through social web communities as they go. If the Millennials are going to start doing anything it's going to be making their very own 90 point wine. They'll want the experience and the wine industry needs to understand that if they do, they will not just want bottles of wine but lots and I mean lots of wine making ingredients and equipment. Millennials are all about experience and what greater experience than to become so informed, you start confidently making your own quality wine year in and year out. They will want to get under the bonnet and start poking around any day soon. You just know they will.

There is just so much prestige in making and branding their own wine that they can resist anything but the temptation of that reality. Including selling some of it into new social media wine communities, in much the same way that farmer markets have sprung up. This millennial wine making movement is about to germinate and when it does wine culture will be radically changed and Millennials will be smiling from ear to ear. They will take back the locus of power and control from the parental industry. Millennials will not settle for just being enthusiasts, they will want to be wine makers too.

A budding Millennial called Shree Bose recently won the Google Kids Science Fair competition for helping to treat ovarian cancer when you have built up resistance to the drugs.[7] She is just 17 and one example of the type of Millennials out there with brains to start making their own quality wine. Dean Kamen, inventor of the Segway, has set up an international organisation called FIRST which runs technology competitions for elementary and high schools to encourage more young people into science and engineering.[8] They had requests from 11,000 students from over 29 countries some of whom came to St Louis to build robots with Lego that even mapped nerves. They will in the future want the whole experience if they can get it, from the soil right through to the glass. Their quest will be to take a global inward looking industry and make it local, using facilities like Crushpad.

What's 'App'ening?

As a result of the success of people now using their phones in the same way as, or even more than, their pc, a new craze called 'apps' has begun and this is also true in the area of wine. Some Japanese bestselling novels have been written entirely on mobile phones. Foodspotting, a food review and photography application, recently crossed the line of 1 million downloads within six months.[9] This, or something like this, could easily be unleashed in the wine universe. Apps can run on your smartphone, meaning phones like the iPhone, or your computer but mostly people use them on smartphones. Apps let you do things like find a restaurant, quickly and without really having to browse which can cost a lot of money on your mobile. Some apps are free to download and can be really helpful. If you have to pay for them they are less than ten bucks.

Take for instance Snooth wine app. You can take a picture of a wine bottle label with your phone and it will direct you to how people have reviewed the wine. The Approach Guides wine app lets you choose and explore wine by style, grape, region and food. You can also get Wine Wherever app which is one of my favourites. This app lets you go around and see the vineyards and will give you the address, phone number and map plus coupons for seasonal promotions.

The wine app in a way is Gary in your hand. The sommelier that you wished could explain the wine that is relevant to you. These apps will help you break through what marketeers call the 'wall of wine'. These apps are the missing link that easily explains the wines to the people who would like to drink them. The apps can direct them to their wine favourites and the apps could even help in getting them trialling different varietals. They are a vehicle to get them the right wine and if wine pairing is your interest, they can be really helpful. More and better wine apps are on the way. By 2020 the apps market will be as significant as the internet. Ilja Laurs, the CEO of GetJar, a company that sells over 60,000 mobile applications, predicts that the smartphone will soon make the desktop obsolete. With global apps downloads soon to go from 9 to 50 billion you can see why apps will be changing how you will make your purchasing decisions.[10]

What all this has to show is the power of wine entrepreneurs, with a lot of talent and in lots of cases not a whole load of money, inventing great services and discovering treasure. The more we have of them, the more the world of wine will become what it should be - totally connected to the enthusiasts. The conversation grand canyon between producers and consumers is radically being reduced in record time and small producers are being given a chance to be heard and seen if they make the effort before the stargate closes and big wine takes over.

Wine 3.0 & Artificial Intelligence

Wine 3.0? Nobody really knows. Your guess is as good as anybody else's, but there are some hints of where possibly this may evolve as regards wine. Data from social media will become the new market intelligence. You could say that wine 2.0 was us creating the web and that 3.0 will be 'self-generating information' when all the data is linked together. Semantically. Currently that is not the case and it's why people are predicting Web 3.0 when it comes. You will be studying, if you are a wine maker, the profiles of your customers more especially their likes and dislikes. Computers will be artificially publishing data in real time.

Your search results will be intelligent and that means that when you put 'cupcakes' into a search engine you will get results personalised to you, rather than 2.5 million results as you now do, and of which you only look at the first page of. The results of the future will include your location, likes and dislikes, profession and education. Possibly also linking you into people of the same interests to converse about similar tastes and topics. Amazon already uses this when it recommends other books from the same genre during your book purchase.

There is a push that all electronic devices will be able to plug into Web 3.0. The robots using artificial intelligence could start twittering thereby creating trends. Let's say a lot of people start a sudden conversation on a new Pinot. The robots will automatically tweet all of the people interested in Pinot wine but who are not aware of the conversation. Within hours there could be a stampede to buy that Pinot. A type of stampede that right now only Oprah's book club recommendations can cause. This is going to be a very exciting development and will be how this information from cyberspace will shape our reality and the wine universe. Big brother may start recommending wine and even wine pairings. Once that data is fused and starts working semantically within a matter of weeks and months - wine marketing will be changed forever. Not to mention our perceptions and our view of the world.

8

The Green Connoisseur

Green. Organic. Biodynamic. Eco-friendly. Sustainable. Biodiverse. This is the new global consumer zeitgeist of wine.

We as wine enthusiasts need to enrol and support this growing wave if we are going to be drinking wines in two decades time. You may be very surprised to hear this but already wine quality and taste is being affected by climate change. Due to rises in temperature there is more sugar in the grape which increases alcohol content in some wines and is denaturing flavour. Temperature is increasing, sea levels are rising and the snow caps are melting - that's the bad news. The good news is that the green eco wines taste great and contain eco terroir made by wine makers who believe passionately in real wines without contributing to the further destruction of our global habitat or global warming. They also don't use pesticides and other chemicals.

What better way to start your entry into the wine universe than through the green door. You can travel the world, partake of some of the best wines - organic, biodynamic and sustainable and there are scientific studies to underline the benefits people accord to organic wines.

There are over 2,000 wines full of taste and story from all over the world to choose from. Choosing to drink green wine made either in an organic or biodynamic way is no longer eco-chic - it's crucial. You don't have to worry about any chemical 'nasties' that could harm you. Many green wines are winning awards. Our global purchasing power could even influence positive measures being taken in agriculture if we support this essential eco sustainable movement in the wine universe. Kofi Annan, ex Secretary General of The United Nations, has publicly supported wineries involved in environmental management in response to climate change.

One of the biggest political actions we take on a daily basis is when we decide what we will eat and drink. Like it or not it has global consequences. Our individual choices, all joined together, shape industries and global economies. With that comes a huge responsibility and modern people understand this with a survival-like awareness. This environmental 'green' drive has been consumer led and is the man and woman on the street's natural concern for his environment and habitat. We just know we just do, instinctively, that by polluting our environment we pollute ourselves.

Some of the chemicals used on vines are highly poisonous and some farmers in France have even died using them. Over 40 sick grape growers there have been connected to pesticides.[1] Yannich Chenet who died from leukaemia from the pesticides he sprayed on his grapes was surprised to find that the people who manufactured the leukaemia medicine were the same company who made the pesticide that killed him. We will see later in this chapter natural and effective alternatives to these pesticides.

So it figures that if we are wine enthusiasts for health we also need to have a keen interest in green wines that are being produced without these toxic substances. The green consumer is eager to learn and is willing to pay up to 20% more for green products that fit their lifestyle and values.[2] These individual choices can have major ramifications globally when you add all the individual choices together. Today over 85% of Americans have bought green goods and are also willing to pay 10% above the standard price for green,[3] even if the packaging alone is just a green 'spin' or what is now called 'greenwashing'. In other words a product that is packaged with green colours but that has no benefit to the environment. A commercial dupe and totally unethical behaviour by ruthless producers.

Wine from The Green Man

The best wine is as natural as possible and was made this way up to thirty years ago. At that point along came modern farming and all its chemicals which dramatically affected the environment where the wines were traditionally grown. Whereas before chemical farming vine roots would grow to a depth of 20 meters, nowadays they won't grow beyond one and grow sideways in order to find nutrients. Vines used to have a life span of a hundred years and today that is now thirty. This is going to affect taste and quality and the way to deal with that, according to oenology, is to have over 400 aromatic yeasts to choose from rather than working with the natural yeast on the grape. Great wine makers are natural wine makers and are very closely connected with terroir and avoid Parkerisation.

So you see the green wine movement is the way wine should and could be and nothing new in some ways. Now it's one of the crucial ways that the industry as whole will survive. When you destroy the biodiversity you need to spray lots of chemicals to protect the vines from susceptibility to disease. Or indeed iatrogenic or man-made diseases from all the chemicals put on the plants and land, as you will have removed natural parts from the ecosystem that manage this in nature. To give you some idea of how deep green goes and how far removed from nature chemical farming is consider the following: Most people don't know that we have killed the soil and therefore the food webs with modern farming methods.

To convert to these organic practices a farm would need to do so over a three year period. Only after the land has been run as an organic farm can it have organic status. In other words they must not use chemicals on their land for that time period. The same goes for a vineyard wishing to go organic.

A ground swell of political action in terms of global warming recently swept the headlines and this was hotly debated scientifically. The Gaia Theory by environmentalist James Lovelock was briefing us three decades ago that we are all connected. As Chief Seattle would say 'like the blood that unites one family'. Whatever we do to the web of life we do to ourselves.' Lovelock described this in terms of a biosphere where ecosystems were in symbiotic balance or exchange harmony. Some political figures in the US are trying to say we are being duped about environmental destruction. Even some scientists. The evidence is all around us of what they are trying to deny. Plants take CO_2 from us and give back oxygen. These exchanges are regulating our whole planet and we cannot live ignoring them. The vineyard biospheres are microcosms of the global biosphere. Everything that happens and what we do within them is a micro-system of the Gaia biosphere.

Along came organic produce which may not have looked as 'zingy' as their pesticide ridden cousins, and modern agriculture did a lot to destroy nature over the following decades. A strawberry in the UK will look great but have no taste whereas one in Spain may not look uniform but will taste fabulous. Spanish people require supermarkets to supply food with great taste. This organic movement slowly developed until the global warming crisis became mainstream and the term 'organic' was pushed to the front. An environmentally aware Vice President rolled out 'An Inconvenient Truth' and a Gaia renaissance took place, creating green consumers overnight with a purchasing power of over $300 billion in just the USA alone.[4] When they realised how they were destroying their natural habitat they jumped to action.

Green consumerism has been a big trend in the retail world. However as I have already said, not all green products are as green as they seem. The industry quickly took advantage of this consumer awareness and now many products that are packaged with green 'spin' are not very green at all. This is something we need to guard against happening in wine to maintain consumer confidence and strengthen the drive for true sustainability.

Consumers are aware of the ecological impact of their purchasing choices. The wine universe has lots to offer them in terms of green: organic, biodynamic, sustainable, climate change wineries and biodiversity. This green movement in wine is not a fad or something superficial in terms of public relations. In fact if the wineries do not become sustainable and are not supported by the eco conscious consumer, they may disappear altogether in a matter of years. This tome is no marketing gimmick at all. As the planet heats and climates change we are already seeing drastic crises in the world of wine. The alcohol levels are rising which causes a loss in taste. California has seen the first frosts ever and over 80% of Australian wineries were recently destroyed with flooding.

The Organic Wine World

When you step into Appellation wine store in Chelsea New York you're walking into the world of green and eco-friendly wine. The bottles are laid out and presented the same way you would expect in any upmarket wine store. The only difference is that these wine bottles have had 'the green touch'. Scott Pactor was an accountant and limelighting wine enthusiast before opening his store which is now a hub for eco-friendly wines and those who are interested. These organic and biodynamically grown wines are fast becoming popular, not only in terms of their meaning, that they are sustainable and ecologically friendly but they taste great and have won a whole raft of medals. The best white wine in the world that has repeatedly won the prize is a biodynamic white wine, Clos de la Coulée de Serrant. Scott is one of the new wine doyens that are emerging in the wine industry and revelling in this growing interest about green wine, commonly referred to as organic.

So what is organic wine and how does it differ?

The term 'organic' that is now in our vernacular basically means natural as possible, no chemicals. In the 1960s there started to be more talk of organic, but it's only in the last decade really that the term has become of major interest to most consumers. This is largely due to the benefits to health the organic produce has, and consumers desire to avoid chemicals that could be carcinogenic. In addition people are concerned about the environment and buying organic produce is seen as a way not to further pollute or do damage. In fairness to some of the viticulturists that produce organic wine they have been in the sector long before it became trendy. They have always known working organically with the soil and the vine produces better wine. Common sense for them. You could also say that organic or normal farming methods have always been there as a natural and indeed traditional and safe form of agriculture.

Organic wines are those that are grown and produced without the use of chemical fertilisers, pesticides, fungicides and herbicides. Organic wine like food is certified, meaning that the vineyard is run in a certain way and is chemical free. The vines will be grown in soil that is alive and will therefore have more mineral content in the fruit. Minerals help the health biochemistry in humans. For example selenium is known to be a preventative agent in terms of cancer. When you choose green wines you're helping your health, protecting the environment and decreasing carbon footprints. Ipso facto.

As conventional vines are sprayed with more pesticides than any other crop this may have serious health consequences. In 2006 the European Commission stated: 'Long-term exposure to pesticides can lead to serious disturbances to the immune system, sexual disorders, cancers, sterility, birth defects, damage to the nervous system and genetic damage'.[5] In South American vineyards you can see the damage to the farmers who have little training to use these chemicals. Their children can be born with deformaties.[6]

To avoid these in what you drink, green wines either organic or biodynamic are your answer.

To start to understand the benefits of organic wine, we first need to look at the benefits of organic food. We know from the UK Food Standards Authority that organic food has apart from a lack of chemical cocktails, 53.7% more beta carotene which protects you against heart disease and cancer, 11% more zinc - a component in healthy immune function, and 38.4% more flavonoids, many of which work as antioxidants.[7] Probably the best person to ask in the world of science is Professor Carlo Leifert who heads the Quality Low Input Food Project at Newcastle University. The project is funded by the European Union with a total budget of €18 million which is a significant amount in terms of scientific research.

Leifert found that organic food had 40% more antioxidants than non-organic food.[8] When his team looked into milk they found that organic milk had 60% more linoleic acid (CLA) and antioxidants than normal milk.[9] Linoleic acid is a popular supplement connected to reduction in body fat for dieters. Organic milk has been known for some time to have higher levels of omega 3 which is good for your brain. The Soil Association issued a report that looked at over 400 scientific studies revealing significant differences between organically and non-organically produced food.[10] So there is a case for organic food scientifically as being better for you. Most people who eat organic food say that it's better in terms of taste and texture. They also believe they are safeguarding their health and wellbeing and they are right.

As regards organic wine we know from a study that Miceli did in 2003 how organic wines compared against standard wines.[11] They looked at 15 wines: five table wines, four Controlled Domination of Origin or DOCs and six organic wines. The organic wine was shown to have the most polyphenols containing 30% more than table wine. Resveratrol and antioxidant activity was also shown to be highest in organic wine. Not only are you getting better tasting wines but also quality in terms of the phytonutrients. One quality control tool that is being developed for organic wine is using infrared (MIR) spectroscopy. Dr Daniel Cozzolino says what he has found is very cost effective. He measured red and white wine samples from over thirteen different Australian wine areas and found the system was correctly able to classify 80% of the samples as organic or otherwise.[12]

By actively supporting biodiversity using different plants, the organic vineyard can tackle unwanted disease and insects as this mix of plants creates a naturally balancing ecosystem. For example, growing certain trees with the vines will attract wasps which will kill the vine leafhoppers that damage vines. Controlled release of ladybirds into the vines controls vine aphids. Mildew is controlled by spraying copper sulphate - the

Bordeaux mixture - on the vines. You may have met this in high school chemistry class and made a blue crystal with it. This has been used as a treatment on vines for over 100 years and the EU is planning a ban. The mix is dangerous and some of the copper finds its way to ground water but this has been the only chemical that has been used in organic viticulture. There are other alternatives that can be used that have been developed in biodynamics. Organic wine growing also increases the existence by up to 50% or more of plants, insects and birds on those farms.

Real organic wine has no chemicals added and is certified - the best being USDA Wine & Vineyard. When a label in the USA says 'organic wine' then it's USDA approved, organic and natural as possible without any chemical or mechanical interference. In Europe it's illegal to label any wine 'organic wine'. If a wine states that it's with 'made with organic grapes' then that is not, let's say, as green, but hey it's still without all those chemicals your body does not need even if the commercial vineyard does. The second definition allows some management using physical processes such as filtration, reverse osmosis, wood chips added to flavour the wine and some clarifiers. So 'made with organic grapes' is still as good as, and even better for you than ingesting pesticides.

The Sulphite Debate – Wine's Deodorant

The biggest contentious issue in terms of green wine is the whole 'to add or not add' sulphites. Some believe these to be the allergy causing chemical in wines. So organic wine is really great for the person who suffers from allergies as the organic wine has minimal sulphite levels.

The addition of sulphites is normally used to kill yeasts and bacteria and preserve wine. All wines produce sulphur dioxide at some level according to Professor Roger Boulton Ph.D. University of California at Davis. A human will produce over 1,000 mg of it a day. We know that 1% of the population is allergic to this compound and the most in danger are asthmatics. For many years it's been blamed for the hangover effect from red wine however there is no scientific evidence as yet to support this. Wine labelled 'organic wine' has no sulphur dioxide added nor does biodynamic wine which is also one of the big attractions. However wines labelled 'made with organic grapes' do. Although they still do not contain the levels that conventional wines have:

- 350 ppm USA standard wine
- 10 ppm Organic wine USDA certified
- 160 ppm EU Red Wine
- 210 ppm EU White wine

- 210 ppm EU Rosé wine
- 150 ppm Wine made from organic grapes
- 100ppm Biodynamic wine Demeter Standard (level allowed but normally much lower)

Sulphur dioxide stops oxygenation and has been a very useful and somewhat natural tool in wine making. It has been used as a preservative for well over a couple of centuries and started by burning candles of the stuff into wooden casks before pouring the wine in to be sealed and shipped. If wine makers don't use it then it's risky business for them as their crop could, during the first stage of fermentation turn to vinegar quite easily. Whites like Chardonnay are drunk quite young but certain reds like Cabernet Sauvignon, as they age, the character comes out. If they oxidise you lose the wine.

When you take a bite of an apple and leave it for a few minutes, where the white flesh turns brown, that is oxidation. This is why winemakers prefer to add sulphur dioxide and why some organic oenologists like to add a little to stop the wine browning and therefore spoiling. This keeps the wine stable and helps it last longer particularly red wine. Too much SO_2 and the wine will smell like a struck match. Too little and the acetaldehyde 'off' odours that it also strips will be found in the aroma of the wine which is lovingly referred to as 'organic funk'. So SO_2 is actually wine's deodorant you could say. In younger reds it will bleach the anthocyanins which are the compounds that give red wine its colour. As the wine ages these then bond with tannins and are not affected by the SO_2. However many wines in France and Italy, if you buy them directly from the locale, won't contain added sulphites.

This may have been the reason that in the early days of organic wines, twenty years ago, they got a bad name. The organic wine makers may not have perfected their art and standards at that time. Wine makers who deal with 'wine made with organic grapes' can however use some sulphur dioxide and this may be helping a great deal due to the amount of awards that these wines are now receiving.

When the European Commission was recently considering banning SO_2, you can get a feel of how important sulphur dioxide is from this following excerpt of an open letter signed by all the European wine lobbying groups.

"Should the use of SO_2 be banned, a great majority of wine-producing units would find themselves in a very difficult situation, since there is currently no substance that could replace SO_2 in disinfecting and preserving wine containers in almost all European winemaking units. Alternative processes are currently being developed, but they appear to be even heavier, since they require special equipment. Therefore, they cannot be used for example by small wineries, which account for a huge majority of European wine-producing units. The use of SO_2 therefore offers, in comparison with those other products, many technical advantages and little impact on the environment."[13]

Organic farming generally improves the environment by first keeping the soils alive. In a 21 year study that took place in Switzerland they found organic farming practices increase life in the soil that conventional farming destroys. They found a number of characteristics:[14]

- Abundant earthworms create 30% higher biomass
- High level of symbionts
- High level of micro-organisms
- 40% more microbial carbon
- More enzymes that help deliver nutrients into plants
- Wild flora that helps pollination
- Energy efficiency by using carbon for growth than maintenance

- Better erosion control because of soil structure

The organic vineyards are also very often green wineries. They create biodiversity by growing different plants, they recycle any water and reduce their energy consumption. Their green winery culture can and very often affects everything they do which massively reduces their impact on the environment. Sabay wine based in Northern Thailand from the Red Bull family use elephants to work their vineyards and harvest the crops. Elephants across Asia are under threat due to having no more traditional use in wood production. Vineyards would be a good place for them to have a life and be protected.

If there is any problem in terms of organic, it has to be the standards, as they can vary from country to country and in Europe wine is not allowed to be even labelled 'organic wine'. Adam Lechmere, Editor of Decanter Magazine, said, "In terms of organic wine we need some form of international recognition that everyone will understand. At the moment organic in New Zealand means something completely different in France. So we need some sort of globally recognised definition of organics."[15] However there is still much to do in getting the wine enthusiasts to go more green and that could even include wines made without pesticides which would be a start and a help to nature. Paolo Bonetti,

President of Organic Vintners in Colorado and wine importer, said that only 0.1% of all wine sold in the US was certified organic.[16] So there is a way to go to supply a huge potential market.

Some wineries, even though they are organic in terms of what they practice do not have or do not want certification. This is possibly due to the costs involved and also the regulations and paper work. They have probably been organic all along, so don't feel they need that status. So there are some wines, not many mind, out there that are organic but are not labelled as such. Even very fine wines with big reputations in Bordeaux are beginning to employ these organic practices such as Chateau La Tour de Bessan Margaux, Lafite, Latour and Jadot in Burgundy. The finest wine in the world, Chateau d'Yquem has a 400 year reputation which is fully organic. So it's good to go to a store that specialises in organic and biodynamic wine as they will know the wines that are made naturally or organically but do not seek certification. There are quite a large number but perhaps you need to be working in the wine world to know.

Vin Au Naturel - La Lutte Raisonnée

Pierre Jancou looks as if he has just walked of the set of a Luc Besson movie, but is a chef restaurateur and devotee of *vin au natural*. When it comes to wine the natural way, he is a bit of an evangelist - a safe and intriguing one. One that demands nothing less than unsulphurated wine which is why he set up his store in Paris to cater for people with the same natural wine appreciation which is called More Than Organic.[17] For the *Vin Au Naturel* movement, some organic wine makers still interfere too much with the wine process. For them natural means just that and absolutely no intervention. There are a number of Parisian restaurants that now focus on these wines. Pierre sources his wine from both France and Italy and insists on unsulphurated wine. "The main secret for making a natural wine without sulphur," he tells us, "is *élevage*, the way the wine is raised." Perhaps also treated in terms of harvest time and also natural fermentation.

This movement was given life and name by the very talented French oenologist Jules Chauvet. In France he is regarded also as one of the best wine tasters ever. In his books he explains exactly how to make natural wine. He was also an expert in Beaujolais and was a friend and colleague of the Nobel Prize winning biochemistry scientist Otto Warburg who he worked with to find a particular microbe for making white wine. As a man with such an amazing scientific mind, oenological experience and family tradition, we have to listen to what he is saying if with all his knowledge he is asking for minimal intervention in wine making to produce quality wines. He would regularly take off to spend time in Grasse - the centre of perfumery - where he practiced with the perfumers to perfect his sense of smell for wine. This man had a passion for natural wines that is beginning to be intriguing to the wine enthusiasts. As the wine is made from low yields, this could also be regarded as an artisan wine.

One wine enthusiast said of one natural wine that it's unlike any Sauvignon you've ever tasted! If you're lucky enough to be in, or passing through, Paris why don't you pop into La Cave De L'Insolite wine store or listen to baroque whilst tasting at Mi-fugue, Mi-raisin and Le Chapeau Melon where Olivier Camus, one of the first pioneers of fine wines resides. If you want to get some insight from Pierre Jancou then head to Vivant his latest natural wine venture.

The Homeopathic Vine

In New Zealand by contrast they successfully use homeopathic preparations to spray on their vines that work very well. Glen Atkinson from GarudaBD told me that, "We use ThermoMax for frost protection, PhotoMax for dull periods from spring onwards, FG4 for support of fungal reduction - but not total fungal control, RipeMax to enhance quality and speed up ripening, E7 - for drought reduction, SilicaMax for stronger plant and flavours. We use ZeroIn to reduce splitting - this last year we achieved no splitting on Pinot Noir grapes at harvest with 100mls of rain." These are all biodynamic homeopathic preparations which are natural and have been scientifically studied. If you need some then head to www.bdmax.co.nz or Glen's company GarudaBD. His preparations could save your life if you are a farmer. The cost of pesticides has increased by over 300% this year making it necessary to look for alternatives.

In the EU over 100 million people use Homeopathic medicine every year. Archanus is a European wine made from grapes with 100% homeopathic use. This is a first and will be marketed to the Portuguese and Brazilian market. If Archanus can do it so many others can. Homeopathic medicine is very popular natural medicine with over 60% of the French population using it and over 100 million people in India. This is a medicine based on energetics rather than chemicals.

One very good use homeopathy has is in a swine flu epidemic where it was shown to be more efficacious than drugs in saving lives in three clinical trials studied during a flu outbreak. The homeopathic remedies Anas Barbariae and Influenzinum were studied and tested and these double-blind, randomised, and placebo-controlled clinical trials were published in the peer-reviewed British Journal of Clinical Pharmacology and British Homeopathic Journal. Over 5 million French people use Anas Barbariae a year. People recovered in 48 hours and the remedies were shown to reduce the effects of flu symptoms by 3 days.[18]

Biospheres, Cow Horns & Stars

Nettle tea for tired vines? Vortex stirring manure? Astrological harvesting? Ground quartz mixed with rainwater in a cow's horn? What's all this? Alchemy you say? Nah, it's just some of the practices in biodynamic agriculture. These wines are produced in the vineyard and not the cellar with the surgery of oenology. Enthusiasts say after years of wine tasting that biodynamic wines have soul. There are scientific studies emerging with some interesting facts about the practice of biodynamic wine. Wine writer Jamie Goode has called biodynamics "a supercharged system of organic farming" in his blog WineAnorak. Jonathan Russo, Publisher of Organic Wine Journal says it best perhaps: "For the very first time when you drink biodynamically grown wine you are not inhaling and not ingesting a cocktail of herbicides, pesticides and fungicides which if anyone would like to tell you are good for you...are f****** insane."[19]

Biodynamics is to wine what Rousseau was to romanticism - 'a return to nature.' Nature does know best and through the practice and the patience of this form of agriculture. Biodynamics then is a return you could say to terroir. Nicolas Joly has got to be the Pan of wine biodynamics. Not only was he practicing this form of agriculture long before it arrived on the map, he was also downsizing from a life as a banker. For over 30 years he has been practicing and developing his biodynamic craft and fighting for 20 years for what he calls 'true wines'. Joly believes he has carved a path for both wine makers and tasters to meet real wines.

Like many burnt out execs and gap year desiring young professionals, after a stint as a banker in New York he decided to move home in his 30's. This happened to be a vineyard in the South of France that many people spend all their lives dreaming of owning. His family and in particular his mother was tending to the vines whilst he was 'making it' in the city. He began by using modern farming methods and saw in one year the soil change as a result of the chemicals he was using. All the ladybirds and partridges disappeared. He felt there had to be another way and then as if by divine providence a book on biodynamics arrived. Back then there were only 5 biodynamic vineyards right across Europe. Now there is a co-operative started by Joly and his compadres that now has 120 members. He accepts people who can produce wine biodynamically with character.

To produce good wine he explains you need to understand the 'sphere' of the vineyard as a whole.

All the living factors, the whole system, is integrated and guided by nature. Joly likes to be known as 'natures assistant'. He likens what he does metaphorically to acupuncture and the Tibetans actually do practice a form of acupuncture like sacred geomancy on the land. He wants to balance the ecosystem to work holistically - as a whole. This is where nature does her best work. To him oenology is a type of cosmetic surgery where wines will lose their natural beauty. This cellar art practice is what he calls 'AC L'Oreal'. More radical than that he believes that modern wine chokes the body.

Joly started biodynamics with his vineyard in 1980 and began to see the results eight years later. In his understanding the core of biodynamics is the antithesis to modern faming which through the use of chemicals, kills the micro flora that enliven the soil with *élan vital* which comes through in his award winning wines. Joly's white wine, Coulée de Serrant, has been voted the best white wine in world by wine critics

Chemical farming kills the soil and the sap. His vineyard dates from 1130 and his vines are quite old by other standards and would have been replaced. His vines of 30 years of age have roots that penetrate 20 meters into the soil in comparison to vines grown with fertilisers which only grown one meter deep. He even has the pictures to prove it and it really makes a mark on you. If true terroir in wine is made from wines with roots 20 meters deep in a natural biosphere, what is the terroir wine critics speak of in commercially grown wines?

If fertilisers are doing this to vines what are they doing to everything else? He won't change his vines for clones which have no character and/or wisdom either. He believes his vines between forty and seventy years old contain a 'vine wisdom' which produces much of his wines many characteristics. Joly et al are also not afraid to lose a batch that fails to ferment for whatever reason. They let it go and don't force something to start that nature does not want. By keeping his ear to the earth and working with the natural heartbeat of Gaia Joly is succeeding in his vocation. Each harvest creates a new wine with a new story - a history of the biosphere that year you could say, reflected in the wine.

Nicolas Joly is a man of wonder and amazement - a true pioneer. His wonder at nature deepens day by day and his amazement stems from the fact biodynamic techniques are not being adopted further than they have, over the past three decades that he and his peers have been using them. Either way he has left a legacy in wine culture and science. He has helped us all look at something from a different angle which clearly works and which science over time will prove more and more to be the case. Some of the ethereal character that Joly produces naturally in his wine is a gift that he would say is from nature itself. To find a few nymphs in his vineyard would not be very surprising. He really puts a symphony in his wines. Joly is a brave wine maker indeed to avoid the Parkerisation of his art.

Astro Tutti Frutti – Supermarket Tastings

Over the past year British supermarkets such as Tesco and M&S have been incorporating one biodynamic wine principal into their diaries - tasting days. You see according to the biodynamic sowing and planting calendar there are days that wine will taste better. These are the days that the supermarkets are using for their tasting flights. In biodynamics the position of the moon affects taste. The moon also affects the sea tides and in Tibetan Medicine, a life energy called 'la' rotates from the foot up the body gradually with the moon cycle. At the full moon it's at the crown chakra of the body. The biodynamic farming calendar is based on practices tried and tested for over 50 years.

Organic wines are easily accessible through wine stores that focus on them and who like Appellation will have tasting notes and helpful young staff who have a contemporary interest in wine. Additionally, you can order them from organic wine specialists by mail order. They offer great and reliable selections and special offers. Even directly from organic and biodynamic vineyards like Frey which will support and encourage the continuance of the green wine movement. When the industry starts to see this sector growing as much as rosé and sparkling wines, we as enthusiasts can start to shape the future of the wine industry to become more green: organic, biodynamic, sustainable, carbon zero and eco-friendly. Doing so we can fill the fields with life again and reduce the amount of pesticides being used in nature.

The greatest challenge to the world of wine now and for the next two decades is climate change. In fact Rudolf Steiner visionary and founder of the Steiner Waldorf education system predicted that there was going to be a problem with bees at this point in time. Bees are now at a point of collapse and are being threatened with extinction. This he saw as due to a problem with pesticides. Without bees we cannot pollinate the flowers and nature will be incapable of producing fruit. This is not the only problem for wines and nature. Climate change is causing problems - huge ones - that threaten the whole industry with destruction.

9

Wine in Climate Change

Climate change is having a huge impact on the planet particularly nature and its ecosystems. In the world of wine this is detrimental and will affect the very existence of wine within two decades.

By then scientists are saying Rieslings will be made in Sweden, red wines in Germany and Champagne production will move to the South of England. In fact sparkling wine production has already begun there. The impact is quite intense and broadens the lens of eco-worry from the initial concern about global warming which is just one card in the pack of climate change. If we can inform wine enthusiasts and create a *cause célèbre* we can affect change and perhaps turn the crisis around before there is no wine left to enjoy.

The extremes of weather change can be seen in the industry globally. Wine terroir is made up of soil type, climate, environment, water levels and even the spirit of the place vines are grown which climate change will affect. Scientists at Stanford have said recently that if temperature increases by 20°F then the Californian wine industry will lose by 50%. Perhaps over half its value which currently is worth $16.5 billion and produces some excellent wines.[1] Noah Diffenbaugh, Assistant Professor at the Woods Institute for the Environment at Stanford, based his study on the assumption that there will be an increase in global temperature by 1.8°F due to a 23% increase in greenhouse gas by 2040.[2] The growers of Pinot Noir will have to start thinking about growing different varietals if temperature increases beyond 68°F which is Pinot's optimal growing temperature.

Nigel Greening founder of Park Avenue Productions - a leading marketing consultancy - is also a New Zealand wine maker. If you are in any way in denial or even minimising the danger of climate change as something which is somewhere way off in the future you need to hear what he has to say. "The main message to get out there is climate 'instability' not warmth. This is what is causing all the destruction we see today. The ocean currents of El Nino and La Nina switch every few years and instability to them can be a massive problem. Australia has seen six years of the worst droughts during El Nino and devastating floods from La Nina. In Victoria eight years ago they were producing Pinot Noir, Chardonnay and Sauvignon Blanc in what was close to a New Zealand climate. Back then the wind normally came from oceans but recently, often the wind comes from the dessert with a 20 degree temperature increase. This caused huge fires, killed hundreds of people and destroyed hundreds of vineyards. Now many are exchanging Pinot vines with Syrah. In Queensland in 2011 an area bigger than Spain and France was flooded while in vineyards in Victoria rains destroyed most of the crops. Anyone who doubts climate change only needs to walk amongst the destruction of some of those vineyards." Nigel owns Felton Road winery in New Zealand.

The Hot Taste of Climate Change

We are all beginning to see the rise in levels of alcohol in wine. This is due to the grapes being produced with more sugar. This increase is also making wine loose its balance according to super experienced oenologist Riccardo Cotarella. According to him two extremely important tannins in wine - vanillina and malvitina - are now at equal levels in wine. In the 1990s malvitina was at levels of 1,200 mg/L and vanillina was at 2,000 mg/L. Today they are at around the 1,600 mg/L mark with malvitina being at a slightly higher level - say 1,650 mg/L. Malvitina is what gives red wine its bitterness and vanillina its sweetness and softness. Wines in the 1990s were more sweet and soft and now due to climate change causing a shift in the tannin levels, wines are a bit bitter.[3]

International wine consultant Riccardo goes on to say that the wine making of twenty years ago was totally different to how we make wine today. This he puts down to an increase in temperature of one degree Celsius. All the climate change has caused the vines to behave totally differently to how they behaved before. People believe it's just a matter of picking grapes earlier to avoid the high sugar levels in them that create high alcohol wines - it's not as easy as that. If they don't wait for the grapes to mature then there will be even lower levels of vanillina. The softer tannin malvitina occurs at the start of the grape growth and vanillina develops at the end. To pick the grapes earlier would cause the wines to be more bitter than they already are.

Two groups have been raising awareness of the impact that climate change is having on wineries globally. 'Wineries for Climate Protection', headed up by Miguel Torres and 'Climate Change & Wine' founded by entrepreneur Pancho Campo MW. Pancho is a retired professional tennis player who threw himself into the world of wine from Dubai where he was in the entertainment industry. After entering the industry he went on to qualify as a Master of Wine. The first conference of his organisation was with an address from Al Gore, ex

Vice President of U.S. Just shows how much an entrepreneur can achieve when they arrive to an industry in a fresh way. Zen mind is beginners mind. They can improve things for everybody and now he is already being proactive in Asia which is bound to be a huge market with his Wine Futures events in Hong Kong. Pancho is another example of how wine benefits from someone with extra skills.

The wine industry needs more people like Campo who come from other industries with extra abilities and commercial awareness at all levels and particularly at the interface between the wine producers and enthusiasts. Both groups are also supported by political heavyweights who have stalwart reputations. To get more headlines and interest I feel we need more of a 'Brangelina' or Spanish pop star like Enrique Iglesias to bring this into peoples' homes. Enrique would appeal to the 30 million Hispanics of the Millennial generation who have greater health risks wine could help prevent. This would also help encourage people to engage with the world of wine if we had these types of Wine Ambassadors that young people can relate to. Throw in Mr. Harry Potter and we are there. Brad Pitt and Angelina Jolie own a biodynamic vineyard in the south of France.

There is an obvious demand for celebrity gossip so why can't some of those column inches be a narrative on wine. This will do more to get the message to the public and make wine and environmentalism a talking point.

Wineries for Climate change have developed a 10 point manifesto for its members.[4] They are:

1. **Reducing Emissions.** Carbon footprint per bottle at the rate set by the European Union 20% in 2020.
2. **Sustainable Building.** Use construction techniques that leverage and decrease the use of natural resources thereby reducing power consumption and overall environmental impact of the habitability of the buildings and achieve integration with the landscape.
3. **Renewable Energy and Energy Efficiency.** Using alternative energy sources to cover part of the thermal requirements of the winery and implement auditing systems to save energy consumption in production.
4. **Sustainable Agriculture and Biodiversity.** Implement cultural practices for the conservation of natural resources and the environment, limit the use of chemicals and encourage the flora, fauna and soil quality.
5. **Reduction of water footprint.** Optimise water use per unit of production by conducting effective and efficient management of the availability of water in agriculture, gardening and production processes.

6. **Ecodesign.** Introduce environmental criteria in the design of product packaging in order to minimise their impact on nature.
7. **Waste reduction**. Reduce the amount of waste generated and implement recycling and recovery of materials. Use the products of the winery as new raw material production.
8. **Efficient distribution.** Minimise the environmental impact of product distribution, using more energy-efficient transportation (rail, vehicles with lower fuel consumption ...) and optimisation of loads and routes.
9. **Research & Innovation.** Develop research projects directed to achieve the reduction of natural resource use, waste generation and emissions of CO_2.
10. **Communication.** Sensitise providers and workers in good environmental practices and to combat climate change.

Torres, who penned this manifesto, is a member of the great Torres family tradition who make wine. He has a deep connection to the environment and understands that to make great wine you need a great environment. Torres and other great winemakers believe this ten point plan is essential for the sustainability of the world of wine.

Ten Green Bottles

One of the biggest problems as regards carbon footprint for the wine industry is glass bottles. In particular the issue of weight which accounts for 40% of the total cost in terms of shipping prices. Nicola Jenkin is the specialist when it comes to the impact wine bottles have on the environment. She worked for WRAP (Waste & Resources Action Programme) which sets about reducing waste and carbon footprint in the UK. Nicola's role is to explore how to do that in wine and as a result she has some very interesting things to say. "Wine produces over 17.5 billion bottles of wine globally which is 8.5 million tonnes of glass which requires 26,000 jumbo jets to ship them and produces 6.6 million tonnes of CO_2. Each bottle weighs 500g. So it doesn't take a lot to figure out that reducing the glass levels in the wine industry will have such a quick impact in terms of a reduction in carbon footprint." Ain't that a fact.

If we were to ship wine rather than fly it and put the wine in Flexitanks we would reduce 30-40% of the CO_2 we use and also cut costs. Each standard shipping container can accommodate 13,000 bottles of wine in glass. If we bottle at delivery and use Flexitanks we can ship 32,000 bottles worth of wine. Many premium wines are now utilising this technology. Waitrose, the UK supermarket, has undertaken for a number of years now to reduce their carbon footprint by shipping in bulk from Chile, the white wine they sell under the label Virtue, in eco tanks of 24,000 L.[5] They are bottled in the UK in 60% recyclable glass. The estimated reduction is 47 tonnes of carbon emissions per year. For each tank that is 32,000 bottles of wine. How eco-bling is that? No more need to ship 16,000 tonnes of glass. The shipping method by sea rather than guzzling jumbo reduces normal costs by 40%, a spokesperson for the supermarket explained, which is a perk they pass onto the consumer. How many other supermarkets are following their lead?

There are a couple of alternatives to glass the immediate being PET plastic. When you compare the weight of a glass wine bottle of 500g against the weight of a PET wine bottle which is 75g then you can see how one can reduce carbon footprint quite dramatically. At the moment standard glass bottles are produced using press and blow machinery but the thickness of the glass is not uniform across the area of the bottle. Here and there will be weak spots. Using new technology such as NNPB (Narrow Neck Press & Blow) one uses less glass as you can incorporate up to 40% PET into the glass and the bottle will not only be lighter but stronger. Good news for Champagne producers who wish to reduce their carbon footprint.

Producers who believe that glass weight has something to do with quality in consumer perception, only need to glance at consumer research reports to see that bottle weight does not even cross their minds. Additionally you reduce any spoilage whilst shipping and can, in terms of health and safety, reduce box weights to be 1 kilo lighter per case for handling. If sparkling wines reduced their weight load then you could reduce the need for 174,000 tonnes of glass. These are quite impacting figures that we as wine enthusiasts can't ignore. Somehow we need to encourage the wine industry to ship wine more this less environmentally impacting way.

WRAP has worked now with 80 wine producers globally to help them in their green initiatives and have saved 30,000 tonnes of glass by working together successfully. Nicola advises wine producers to use green glass rather than clear flint types, if they can, as the former is 75% recyclable. As regards PET she advises to use clear as that can go straight to be recycled into a water bottle whereas coloured PET goes into the making of seat belts etc.[6]

Wine in the UK produces 600,000 tonnes of waste and would you believe it, over 44,000 litres are thrown away worth £430 million sterling, nearly one billion dollars[7] Once wine is opened if it's not used it will spoil so it is also perishable. More emphasis must be placed on wine storage and packaging sizes and types. This is where we are yet to see lots of innovation and transformation of the wine world which would endear maybe more enthusiasts and trial purchases. People may like to explore wine in perhaps one glass serving, in packaging that does not expensively, foolishly and unnecessarily need to be shipped in a glass bottle on a jumbo jet in this day and age, contributing to climate change which could further contribute to the destruction of the world of wine.

Freshcase is a new BiB (bag-in-box) with innovative technology. This box easily sits in your fridge and keeps premium wine fresh for up to six weeks after being opened.[8] No air gets into the wine when you pour a glass. Over 50% of Scandinavians prefer to have their wine in this BiB form.[9] This Freshcase system holds three bottles worth of wine or 2.5 litres whilst only taking up one bottle's worth of shelf space in your fridge. In comparison to the three bottles it would normally need to deliver the wine this system is 70% lighter which gives it a very low carbon footprint. Through this system you can have top quality wine on tap at all times. The whole thing is totally recyclable and more wine could be packaged this way.

Greenbottle have just released their version of a paper wine bottle which is totally biodegradable and also compostable.[10] Like BiB there is an aluminium lining which some feel is a disadvantage in the paper bottle, as in BiB you can separate the paper box from the bag containing wine. The company who invented the bottle have received a patent for the invention. The weight is 55g so you can see how much an environmental impact it's going to make against 500g of glass. These have previously worked well with milk at ASDA. This could appeal to a whole glut of green consumers and therefore would be a good packaging concept for organic wine. The paper bottle could help organic wine stand out in their own special form in green packaging. You would have read earlier how much the green consumer is worth. This could also help boost sales of organic wines as they would be easily recognisable. So many are intimidated by the 'wine wall' so let's make it easier for people to buy green wine by putting the wine in these eco paper bottles. Stating how much glass wastage is reduced by the purchase could also be an incentive. This could be the future of wine packaging if we could get used to it. Once again I believe we as enthusiasts are ready and it's the industry that is stalling.

Carducci - The 1st Carbon Zero Winery

When Parducci went green they went the whole hog. There was no skimming the surface they went for the full plunge. Today their winery has been independently assessed as the first carbon neutral winery in the USA.[11] This was deduced by analysing their carbon footprint which refers to the total amount of Greenhouse Gas (GHG) that the company or person produces. The winery then set about reducing this in all areas to remedy their impact on the environment.

Seeing that their energy use made up over 60% of their carbon footprint they looked at addressing this first. They resolved most of this by green energy and introducing solar power, bio-diesel and making their operation energy efficient. They then got carbon credits to improve their green rating by financially supporting green initiatives in their local area such as forestry. This carbon neutral policy of Parducci is another success and is equivalent to taking over 80 cars off the road annually. Imagine how many cars could be taken off the road if most wineries adopted this eco policy.

Going green also makes economic sense in terms of the amount of savings people make by becoming more efficient and by offsetting the damage being done by GHG. When Professor Leo Pyle, my professor, who was an astounding genius of biotechnology and biochemical engineering was once asked to see how he would improve a dairy - he saved them £1 million sterling a year.[12] He examined all their waste and he was also a great supporter of the use of bioethanol. He had us design a total bioethanol facility from top to bottom. Companies and the wine industry, with the implementation of aquatic ecosystems as living machines don't have to have their waste water polluting rivers. There are so many solutions.

The Future of Cork Is ZORK

No it's not a strange world in science fiction full of 'zorks' but rather the new future of wine sealants. The inventor of the ZORK is John Brooks and he invented the product over ten years ago when he saw an opening. Excuse the pun. The ZORK pops like a normal wine cork on opening but it's more reliable and safer. After opening it can be placed back on the bottle keeping the wine fresh by being resealable. This could help stop the wine wastage of 44,000 litres in the UK. The ZORK is recyclable and sustainable. For small wineries there is a huge saving. When they use the ZORK as closures for their wine they expend $250 against standard cork and machinery of $15,000.

When you consider what the Tetra pack people are now worth innovation in this area of the wine industry can reap huge rewards for a clever entrepreneurial inventor. Some are predicting that 30% of all corks will be ZORK very soon. The biggest cork producer in Portugal, Amorim, produces over 3.2 billion units annually for wine bottles. So which Caribbean Island would you like John?

The €400 Million EU Wine Lakes

Every year normally the European Commission spends over €400 million from a €1.3 billion wine budget distilling alcohol from wine that fails to get to market.[14] In addition there is finance for 'emergency distillation' depending on the yield each year. Rather than spend that money getting the quality standard of the wines up and to market it's been spent on emergency distillation which is just not sustainable. 2011 was the first year Europe did not create this wine lake which is a huge success for the European Commission.

Most of this wine in France was produced in the Languedoc-Roussillon region, one of the world's biggest wine production regions generating over a third of all France's wine. As you can imagine this has caused a lot of consternation in Brussels, the HQ of the European Commission. Then if you add to this a culture of green (the EU has legislation in place that by 2020 Europe is to generate 20% of all its energy from renewable sources) and the plain common sense mismanagement of the problem something has to be done to find a remedy and make better use of the funds for wine. Thankfully this is beginning to happen.

The EU through its public tender programme traditionally sought and awarded a contract for the distillation of 700,000 hectolitres some of which has been in storage:

- 300,000 France
- 200,000 Spain
- 200,000 Italy

- 60,000 Greece

As it stood the EU was paying some farmers to produce bad wine that needs to be disposed of. These small holdings survived on the basis of the financial subsidy from the EU. What would be good to see is a report of the quality of the wine. Is it really that bad? Or is it a failure of marketing that it fails to get sold. Either way the lake has finally been drained.

Due to the labelling restrictions on wine would you believe that Australian wines do better in France than French wines? The reason is that French wines are branded by region not by the grape in order to comply with EU wine label legislation. Whereas the Australians are making it easier for consumers, who after all have the end say, to choose Australian over French. This is only because of label restrictions on French wines stopping them pushing the grape variety. Interesting and ironic also to note is that the whole surface area of Australian wine is roughly 400,000 acres of vines - the same amount that the Commission wants to uproot in France as we shall see later.

High end fine wine producers are going outside the EU zone to produce wines for a global market without the same restrictions they find themselves under in the EU. Just goes to show as regards entropy - energy flows to the place of least resistance. You can be sure that in the geographic areas that these producers are now based, there is no surplus wine. They just don't have the same bureaucratic restrictions bearing down on them.

Apart from the economic impact of emergency distillation what is the effect to the environment? Sure, we take alcohol out for bioethanol but what about the huge waste stream? Where does that go? What river or landfill? Wine-distillery wastewater has a high organic pollutant load (40g COD/litre) which includes various phenolic compounds, the major ones being gallic acid, p-coumaric acid and gentisic acid. As these polyphenolic compounds have strong anti-microbial effects it makes it difficult to treat this type of wastewater with bioremediation.

Using biotechnology the health compounds of the wine could be made into a product using downstream processing techniques such as reverse osmosis rather than being dumped. The waste stream could also be used to produce biogas. One of the solutions to stop the crisis was to immediately pull up 1 million hectares of vines and plant a different crop. The reaction was uproar by the farmers in France, but 400,000 hectares will still be pulled. The distillation subsidy is to be quickly phased out. When you consider that the total vine area of, let's say, Italy is roughly 2 million acres then you see how devastating this is. More so when you know that the Commission forced a vine slashing of 1 million acres for the same reason in 1998.

So each year the EU was making a quantity of wine that they fail to bring to market. Yet at the same time wine lobby groups are very confident about campaigns of 'moderation' in wine drinking when wine is already a culture of moderation. Take for instance the UK market: wine is generally drunk there only once a month. How much of the wine lake that is disposed of every year - costing €400 million - could we reduce by encouraging wine to be drunk once a week and especially for people who are overweight. Not only could we reduce the wine lake we may also reduce the effects of syndrome X or metabolic syndrome.

The European Commission like the wine industry seems to also be out of touch with the EU wine enthusiasts as they plan to spend just €3 million of a marketing budget on marketing wine in EU countries and over €120 million outside the Eurozone in places like the USA. The EU has a population of over 400 million citizens for a wine market. Wine drinking by these people could significantly reduce the healthcare bill. Joaquim Madeira, President of the ANDOVI, stated "the smallest budget is allocated to the internal market which represents 67% of the world market."[15] Why is this excess wine not sold in the new emerging markets like India and China? The Commission needs to do more.

The other problem with the new measures is that grape growers in the regions who are in financial crisis will be forced to sell their land to the big wineries. This will have a devastating impact both in terms of personal crisis of farmers who have grown grapes all their lives and inherited the tradition for generations. Their knowledge will be lost and this will cause rural depopulation. Some of the farmers are suffering suicide risk due to all the problems - another good reason to sort the measure out and get the area producing better quality wine. This could easily be exported to China and Russia who are both promoting wine consumption in favour of spirits for the social benefits.

Ten litres of wine will yield one litre of pure ethanol and the biggest problem is trying to get the ethanol out of a litre of wine which is 90% water. Perhaps that wine could also be repackaged and blended for hospitals and even old folks homes. Let's make it easier for these wines to come to market where they can benefit people rather ending up at the distillery where the wine is used to make bioethanol which will use up a lot of energy getting the alcohol out of the water.

Is it that the wine is of poor quality or the marketing and distribution is failing? Why can't we supply wine to all the hospitals with this material or send it instead to the 400 million people in the EU. European wines have to be marketed in Europe as well as the USA.

Solar Power Wineries

Solar power panels work by harnessing the sun's light using solid photovoltaic cells that convert this light into electricity. They were first popular in fuelling satellites starting with Vanguard I, the fourth satellite ever in space. Constellation Brands USA have put 17,000 solar panels in four wineries that combined give a 3.5 megawatt DC power source which removes 4.5 million pounds of carbon from their carbon footprint.[16] Their spokesperson said, "Better still it's not producing exhaust fumes from a car driven 226 million miles over 25 years." The solar power produced will create half the electricity needs of Constellation's Gonzales winery in California. When the winery does not need electricity during the year it sends it to 1,000 local homes. Although thieves are cashing in. They are using Google Earth to locate which vineyards have the solar panels in California.[17] Quite lucrative criminal activity with each panel being worth $1,000. If they find 100 panels – well do the maths?

Scientists at MIT recently discovered that a bacteria eating virus called M13 could moonlight as a 'nanoworker' for solar power cell production. Professor Angela Belcher at MIT saw that the viruses could make sure the nano components work and raise efficiency of the cells by a whopping 30%.[18] They had been investigating carbon nanotubes, which increase the rate of electron collection from the surface of the solar cell when they came upon the idea of using biology and physics. Not only do the viruses keep the nanotubes in place but also create a layer of titanium dioxide, a key component in these solar cells, rather than the conventional ones made of silicon.

Biogas from Waste Grape Skins

Inniskillin Winery in Canada teamed up with StormFisher Biogas in Ontario to turn grape pulp waste into biogas.[19] The anaerobic digestor uses all the skin and seed grape waste produced by the winery - up to 2,000 tons. This would normally go to landfill and the greenhouse gas methane is now being used to fuel electricity in some homes rather than being added to the atmosphere. As the EU strives to reach its 2020 green goals this is a very interesting scheme for all European wineries to become involved in. Grape waste sent to biogas will stop methane being used. Methane is twenty times more harmful than carbon dioxide to the environment. Dumping to landfill, apart from being detrimental to the environment and unsustainable as these sites are now very limited is also extremely expensive. Dispatching the waste to biogas centres is relatively cheap. Each StormFisher biogas plant can digest 120,000 tons of organic waste a year and produce 2.5 megawatts of electricity which can supply 2,500 homes. Ontario produces 13,000 tons of grape waste a year during wine making. In Europe in 2000 there were 300 digestors and today there are over 4,000 but more needs to be specifically dedicated to wine making.

Geothermal Energy - Warmth from the Earth

At levels of 15 meters deep the temperature of the earth remains stable through the year in certain parts of the world. This creates the case for harnessing what is called geothermal energy. This type of energy is independent of weather fluctuations that both solar and wind are affected by. An ambitious project, a first, funded by the government of the Rioja region La Bodgea Institucional Grajera which is part of the ICVV has now been completed in Spain. The construct is a modern research centre that is sustainably designed in particular with geothermal energy in mind. The heating and cooling system, HVAC system, has been designed to be 100% geothermal by architects of the project, Arquitelia.

This system can be used to cool fermentations and keep the cellar temperature stable all year round. In the cellar and institutional building the geothermal heat exchanger consists of 40 holes of 127 feet deep each, with an installed capacity of 420 kilowatts. The administration building was built with a geothermal heat exchanger consisting of 15 holes, each 112 feet deep, with an installed capacity of 65 kilowatts.[20]

Architects of the project explained that this will be a first in the world with a comprehensive geothermal utilisation. This enables energy savings of 68% on what a traditional heating system would use. As regards the fermentation areas the climate control system design allows different temperatures in the winery and institutional buildings. In addition, they have individual temperature control in 26 winemaking areas.

Biochar Terra Preta & the GHG Solution

Biochar was rediscovered in 1950 in the Amazon by a Dutch soil scientist called Wim Sombroek. Some geologists estimate that 10% of the Amazon is this biochar rich soil called *terra preta* or 'black land'. Sombroek was talking about the benefit these soils could have against global warming in the early 1990's. Amazonian soils on their own are not so fertile and the tribespeople knew this. They would burn the biomass covering the fire with soil to create biochar which they then spread. A recent study carried out by Bruno Glaser and featured by the BBC showed that soils with biochar produced 880% higher yields than those soils without and that used fertiliser.[21]

Today biochar in Europe is made by pyrolysis which creates an oxygen absent burning of biomass. Antonio Moran from The University of Leon is investigating biochar for its use in soil. "There is a gas, liquid and solid fractions. As a result of the process the solid being the biochar which is very high in carbon. Main thing is that biochar takes hundreds of years to biodegrade." This traps CO_2 which is a greenhouse gas contributing to global warming.

The carbon structure also changes from the biomass state of lignins to more amphorous and tubostratic compounds. The biomass is not turned to ash like traditional burning where oxygen is more available. Some traditional farmers have for centuries burned wood and then poured water extinguishing the fire before the charcoal turned to ash. Biochar is 70 times higher in carbon than normal soils and is nutrient rich. Some soil scientists believe that it acts as a sponge for water, nutrients and fertiliser. In fact biochar may also stop nutrients being washed away as they are in lifeless soil. If we are to use biochar on all land we can sequester all CO_2 in the atmosphere. Biochar can be made in your backyard or in a pilot plant scale. Hans-Peter Schmidt and Delinat organic wine distributors in Switzerland have devised a piece of kit that each winery can use to produce biochar.[22]

Without any further ado all wineries should have biochar soil and perhaps a notice placed on the labels 'wine made from grapes grown on vines in biochar'. This could have a huge impact on the environment and rapidly reduce the CO_2 levels wineries produce. Dr. Johannes Lehmann, a Cornell

University geologist says, "Biochar can be used to address some of the most urgent environmental problems of our time - soil degradation, food insecurity, water pollution from agrichemicals and climate change."[23] Lehman of all people should know - he has spent years studying the area and knows we could reduce CO_2 emissions by at least 12% from biomass as biochar if we used it.

In order for the world of wine to survive, it needs to come to terms with many realities not least where wine comes from - the environment. We are set to see a global temperature one degree increase. If the oceans rise by just 4 meters Bangladesh and parts of other countries will be under water. More famine and droughts are on their way. What could happen if we encourage African countries to also use biochar? Out of this struggle for survival we will see a raft of new innovations and a sober attitude to carbon emissions. Vineyards can be the place to show a leadership role in agriculture and the environment causing biochar to be spread on as much farm land as possible. This will create a demand for biomass as biochar which will then cause the creation of specialised companies. They can take the landfill waste and use it to make the environment come alive again with living soil. Through biodiversity we can protect the vineyards from becoming vulnerable to biological disasters due to susceptibility caused through a lack of nature not being allowed to be simply nature.

My hope is that the wine industry can lead the way for all of agriculture and be the shining star by caring properly for the vine and in doing so, the environment or Pachamama. By implementing these ecological procedures we may not just save wine but our whole planet which is fast coming under threat from mindless profiteering industrialisation that is disconnected from the earth and the reality of global warming and the destruction of climate change.

Notes

Chapter 1 - The Wine Mind

1. Lockshin, L, W Jarvis, F d'Hauteville, and J P Perrouty (2006), "Using simulations from discrete choice experiments to measure consumer sensitivity to brand, region, price, and awards in wine choice," Food Quality and Preference, 17, 166-78.
2. Revue Vinicole Internationale - March 2011
3. www.winetagefund.com/wine_market_why_invests.htm
4. www.roymorgan.com/resources/pdf/papers/20041201.pdf
5. www.guardian.co.uk/lifeandstyle/2009/apr/02/wine-sales-women
6. www.winemarketcouncil.com/research_summary.asp
7. www.svb.com/2011-wine-report
8. www.winesymposium.com/presentations/2011WIFS/Gillespie.pdf
9. Cohen, D. L. (2011), Entrepreneurs target millennial wine drinkers, Reuters U.S. Edition January 19th
10. Adams Wine Handbook 2006
11. Ritchie, C. (2007). Beyond drinking: The role of wine in the life of the UK consumer. International Journal of Consumer Studies, 31, 534 - 540.

Chapter 2 – The Tasting Lab

1. www.sandhiwines.com/Rajat-Parr.aspx
2. www.tastingscience.info/delwiche.htm
3. www.nzsvo.org.nz/news_&_events/events/2005_sensory_workshop/ parr_published_papers.htm?xid=&xtrkm=0
4. www.rsc.org/Publishing/ChemScience/Volume/2008/09/The_wine_fraud_detective.asp
5. Demystifying wine expertise: Olfactory threshold, perceptual skill, and semantic memory in expert and novice wine judges. Chem. Senses 27, 747–755. Parra et al
6. www.msnbc.msn.com/id/14610793/ns/technology_and_science-innovation/t/researchers-unveil-winebot/

7. Castriota-Scanderbeg, A and Hagberg, GE and Cerasa, A and Committeri, G and Galati, G and Patria, F and Pitzalis, S and Caltagirone, C and Frackowiak, R (2005) The appreciation of wine by sommeliers: a functional magnetic resonance study of sensory integration. NeuroImage , 25 (2) 570 - 578.
8. www.independent.co.uk/life-style/health-and-families/health-news/taxi-drivers-knowledge-helps-their-brains-grow-428834.html
9. www.reuters.com/article/2010/06/01/us-luxury-summit-taittinger-idUSTRE6505 KM20100601
10. Mondaini N, Cai T, Gontero P, Gavazzi A, Lombardi G, Boddi V, and Bartoletti R. Regular moderate intake of red wine is linked to a better women's sexual health. J Sex Med 2009;6:2772–2777.

Chapter 3 - Soil to the Glass

1. www.yorkvillecellars.com/organic_ourphilosphy.php
2. What Is Biodynamic Wine by Nicolas Joly
3. www.carboncommentary.com/2009/05/04/594
4. http://eusoils.jrc.ec.europa.eu/library/themes/biodiversity/
5. www.potashcorp.com/media/POT_tfi_brochure.pdf
6. http://online.wsj.com/article/SB10001424052748703794104575546430039 9 44898.html

Chapter 4 - Wine Transfusion

1. The French Paradox and Beyond: Living Longer with Wine and The Mediterranean By Lewis Perdue, Keith I. Marton, Wells Shoemaker
2. www.nytimes.com/1991/12/25/garden/wine-talk-425591.html?pagewanted=all&src=pm
3. www.ornishspectrum.com/
4. www.ulsterhistory.co.uk/samuelblack.htm
5. www.olemiss.edu/orgs/AWARE/inews1.html
6. The Fat Fallacy : The French Diet Secrets to Permanent Weight Loss by Will Clower
7. http://edition.cnn.com/2009/HEALTH/12/08/breast.cancer.soy/index.html
8. Ziegler RG, Hoover RN, Pike MC, Hildesheim A, Nomura AM, West DW, Wu-Williams AH, Kolonel LN, Horn-Ross PL, Rosenthal JF, Hyer MB.

Migration patterns and breast cancer risk in Asian-American women. J Natl Cancer Inst. 1993 Nov 17;85(22):1819-27. PubMed PMID: 8230262.

9. www.herbcompanion.com/health/fresh-clips-herbal-wines-in-ancient-egypt .aspx
10. D. & P. Kladstrup Champagne Harper Collins Publisher ISBN 0060737921
11. Wine is the Best Medicine - Dr E A Maury
12. Klein and Pitman Journal of Substance Abuse; (Vol. 2, No. 3, 1990)
13. http://news.bbc.co.uk/2/hi/business/369684.stm
14. www.modern-wine-cellar.com/blog/tag/greek-god-of-wine/
15. www.telegraph.co.uk/science/science-news/7863447/Money-cannot-buy-total-happiness.html

Chapter 5 - Wine as Preventative Medicine

1. Howitz KT, Bitterman KJ, Cohen HY, Lamming DW, Lavu S, Wood JG, Zipkin RE, Chung P, Kisielewski A, Zhang LL, Scherer B, Sinclair DA. "Small molecule activators of sirtuins extend Saccharomyces cerevisiae lifespan". Nature. 2003 Sep 11;425(6954):191-6. Epub 2003 Aug 24.
2. www.nytimes.com/2011/09/22/science/22longevity.html
3. http://money.cnn.com/2007/01/18/magazines/fortune/Live_forever.fortune/index.htm
4. www.sirtrispharma.com/
5. http://circ.ahajournals.org/content/111/2/e10.full
6. Effects of vitamin antioxidant supplementation on cell kinetics of patients with adenomatous polyps. R J Cahill, K R O'Sullivan, P M Mathias, S Beattie, H Hamilton, C O'Morain. Gut. 1993 July; 34(7): 963–967.
7. The Red Wine Diet by Rodger Corder
8. Arthur Marx and Raymond R. Neutra - Magnesium in Drinking Water and Ischemic Heart Disease - Epidemiol Rev Vol. 19, No. 2, 1997
9. http://health.usnews.com/health-news/family-health/heart/articles/2011/03/04/ potassium-rich-foods-may-cut-stroke-heart-disease-risk
10. Nature. 2006 Nov 30;444(7119):566. Oenology: Red Wine Procyanidins And Vascular Health. Corder R, Mullen W, Khan NQ, Marks SC, Wood EG, Carrier MJ, Crozier A.

11. Hertog - The Lancet, Volume 342, Issue 8878, Pages 1007 - 1011, 23 October 1993 Dietary antioxidant flavonoids and risk of coronary heart disease: the Zutphen Elderly Study
12. Stampfer MJ et al A prospective study of moderate alcohol drinking and risk of diabetes in women. Am J Epidemiol 1988;128:549-58.
13. Cell Biochem Funct. 2006 Jul-Aug;24(4):291-8. A central role of eNOS in the protective effect of wine against metabolic syndrome. Leighton F, Miranda-Rottmann S, Urquiaga I.
14. Red wine: A source of potent ligands for peroxisome proliferator-activated receptor. Alfred Zoechling, Falk Liebner and Alois Jungbauer, Food Funct., 2011, 2, 28
15. www.dailymail.co.uk/news/article-487122/Red-wine-prevent-food-poisoning-stomach-ulcers.html
16. Kubo A, et al "Alcohol types and sociodemographic characteristics as risk factors for Barrett's esophagus" Gastroenterology 2009; 136: 806-815.
17. www.sciencedaily.com/releases/2008/07/080707081848.htm
18. www.health.harvard.edu/press_releases/prostate-benefits-from-red-wine
19. http://lpi.oregonstate.edu/infocenter/phytochemicals/resveratrol/resverarefs.html#
20. PADMA-28, a traditional Tibetan herbal preparation, blocks cellular responses to bFGF and IGF-I. Inflammopharmacology. 2004 ;12:373-89 15901415
21. Nutr Cancer. 2001;40(1):74-7.Larry Clark's legacy: randomized controlled, selenium-based prostate cancer chemoprevention trials. Marshall JR.
22. Int J Cancer. 2005 Jan 1;113(1):133-40. Alcohol consumption and risk of prostate cancer in middle-aged men. Schoonen WM, Salinas CA, Kiemeney LA, Stanford JL.
23. www.escardio.org/congresses/europrevent2006/Documents/110506EuroPrevent_Session2_Trichopoulou.pdf
24. www.winespectator.com/webfeature/show/id/Some-Red-Wines-Help-Kill-Foodborne-Pathogens-Study-Finds_3869
25. Teyssen, S., Lenzing, T., González-Calero, G., et al, "Alcoholic beverages produced by alcoholic fermentation but not by distillation are powerful stimulants of gastric acid secretion in humans," Gut Jan. 1997; 40(1):49–56
26. www.dailymail.co.uk/health/article-1189949/A-glass-wine-day-cut-gallstone-risk.html

27. Hepatology. 2008 Jun;47(6):1947-54. Modest wine drinking and decreased prevalence of suspected nonalcoholic fatty liver disease. Dunn W, Xu R, Schwimmer JB
28. European Journal of Clinical Nutrition (2011) 65, 526–532; doi:10.1038/ejcn. 2011.9; published online 16 February 2011 Beverage-specific alcohol intake and bone loss in older men and women: a longitudinal study J Yin1, T Winzenberg1, S Quinn1, G Giles2 and G Jones1
29. Ganry O, Baudoin C, Fardellone P: Effect of alcohol intake on bone mineral density in elderly women: The EPIDOS Study. Epidemiologie de l'Osteoporose. Am J Epidemiol 2000; 151:773-780
30. Wang et al. Alcohol Consumption, Weight Gain, and Risk of Becoming Overweight in Middle-aged and Older Women. Archives of Internal Medicine, 2010; 170 (5): 453 DOI: 10.1001/archinternmed.2009.527
31. www.foodprofessionals.org.au/food-companies-corporate-responsibility-and-health
32. PRATT, D. E., POWERS, J. J. and SOMAATMADJA, D. (1960), ANTHOCYANINS. I. THE INFLUENCE OF STRAWBERRY AND GRAPE ANTHOCYANINS ON THE GROWTH OF CERTAIN BACTERIA . Journal of Food Science, 25: 26–32. doi: 10.1111/j.1365-2621.1960.tb17932.x
33. David Vauzour, Emily J. Houseman, Trevor W. George, Giulia Corona1, Roselyne Garnotel, Kim G. Jackson, Christelle Sellier, Philippe Gillery, Orla B. Kennedy, Julie A. Lovegrove and Jeremy P. E. Spencer. Moderate Champagne consumption promotes an acute improvement in acute endothelial-independent vascular function in healthy human volunteers. British Journal of Nutrition, 2009; DOI:10.1017/S0007114509992959
34. J Agric Food Chem. 2007 Apr 18;55(8):2854-60. Epub 2007 Mar 24. Champagne wine polyphenols protect primary cortical neurons against peroxynitrite-induced injury. Vauzour D
35. Vauzour, D., Vafeiadou, K., Corona, G., Pollard, S.E., Tzounis, X. and SPENCER J.P.E. (2007). Champagne wine polyphenols protect primary cortical neurons against peroxynitrite-induced injury. J. Agric. Food Chem. 55, 2854-2860.
36. Boyer, J.C., Bancel, E., Perray, P.F., et al, "Effect of Champagne compared to still white wine on peripheral neurotransmitter concentrations," Int. J. Vitam. Nutr. Res. Sept. 2004;74(5):321-8
37. VinsonJ and Hontz B Phenol Antioxidant index: Journal Agricultural Food Chemistry 1995,3.

38. Jung et al Herz/Kreisl,31 (1/99)pge 25-31.
39. www.goodtasteinternational.com/main.php?id=346
40. GLIDEWELL, S. M., DEIGHTON, N., GOODMAN, B. A., TROUP, G. J., HUTTON, D. R., HEWITT, D. G. and HUNTER, C. R. (1995), Free radical scavenging abilities of beverages. International Journal of Food Science & Technology, 30: 535–537. doi: 10.1111/j.1365-2621.1995.tb01400.x
41. www.sciencedaily.com/videos/2005/1204-wine_cleaner.htm
42. Protective Effect of Structurally Diverse Grape Procyanidin Fractions against UV-Induced Cell Damage and Death J. Agric. Food Chem., 2011, 59 (9), pp 4489–4495 DOI: 10.1021/jf103692a

Chapter 6 - Artificial Intelligence Vineyard

1. www.rps.psu.edu/0009/machine.html
2. www.icv.fr/index.php?module=oenoview
3. www.winespectator.com/webfeature/show/id/Winemakers-Need-Space-to-Make-Good-Wines_4325
4. www.bbc.co.uk/programmes/b00qbbwd
5. www.winebusiness.com/wbm/?go=getArticle&dataId=46234
6. www.fruitionsciences.com
7. http://spectrum.ieee.org/automaton/robotics/industrial-robots/will-robots-pick-your-grapes-one-day
8. www.economist.com/node/15048711
9. http://mars.jpl.nasa.gov/programmissions/
10. www.youtube.com/watch?v=EcXhOsaXKVo

Chapter 7 - SEO Wine 2.0

1. www.vintank.com/
2. www.svb.com/2011-wine-report-pdf
3. www.businessweek.com/magazine/content/11_21/b4229022111543.htm
4. www.snooth.com/
5. http://mysocialwinery.com/
6. www.cellartracker.com/newsletter/1-31-2011.htm
7. http://newswatch.nationalgeographic.com/2011/07/12/google-science-fair-winners-announced/

8. www.usfirst.org/
9. http://thenextweb.com/apps/2011/08/10/foodspottings-app-hits-1-million-downloads-and-gets-new-social-features/
10. www.mobilemarketingwatch.com/forecast-mobile-apps-to-soar-to-50-billion-downloads-by-2012-set-to-outsell-cds-5747/

Chapter 8 - The Green Connoisseur

1. winenow.biz/2011/01/29/pesticides-kill-french-winegrower/
2. www.bizreport.com/2011/09/nielsen-20-of-consumers-will-pay-more-for-eco-friendly-produ.html
3. www.bizreport.com/2011/09/nielsen-20-of-consumers-will-pay-more-for-eco-friendly-produ.html
4. www.marketplace.org/topics/business/buying-green-not-planet
5. http://europa.eu/rapid/pressReleasesAction.do?reference=MEMO/06/278&format=HTML&aged=0&language=EN&guiLanguage=en
6. www.ejfoundation.org/pdf/whats_your_poison.pdf
7. www.i-sis.org.uk/FSAorganicFoodBetter.php
8. www.qlif.org/qlifnews/april05/con0.html
9. www.ota.com/organic/benefits/nutrition.html
10. www.ota.com/organic/benefits/nutrition.html
11. Miceli, A. et al. 2003. Polyphenols, Resveratrol, antioxidant activity and Ochratoxin A contamination in red table wines, Controlled Denomination of Origin (DOC) wines and wines obtained from organic farming. Journal of Wine Research, 14(203): 115-120.
12. www.bfa.com.au/Portals/0/BFAFiles/PDFs/ACOM5-lo_006.pdf
13. www.zdvvs.si/data/upload/Courrier_SO2_Biocide_030511_EN.pdf
14. www.fao.org/DOCREP/005/AD090E/AD090E00.HTM
15. www.decanter.com/news/wine-news/529559/biodynamics-should-be-promoted-debate-audience-decides
16. www.ams.usda.gov/AMSv1.0/getfile?dDocName=STELPRDC5091169&acct=nosb
17. www.morethanorganic.com/
18. The American Institute of Homeopathy Handbook for Parents
19. www.organicwinejournal.com/index.php/2011/03/why-isnt-more-wine-organic-2/

Chapter 9 - Wine in Climate Change

1. Climate adaptation wedges: a case study of premium wine in the western United States, Noah S Diffenbaugh et al 2011 Environ. Res. Lett. 6 024024
2. http://news.stanford.edu/news/2011/june/wines-global-warming-063011.html
3. www.climatechangeandwine.com/videos.php?id=38
4. www.wineriesforclimateprotection.com
5. www.thegreenwineguide.com/2009/06/07/waitroses-virtue-shipped-in-eco-friendly-tanks/
6. www.wrap.org.uk/downloads/Drinkspack_11_Presentation_-_Nicola_Jenkin.ea95c770.10703.pdf
7. http://cambioclimaticoyvino.com/conferencias/conf8/8_1.pdf
8. www.freshcasewine.com/default.aspx
9. www.winesur.com/news/bag-in-box-exports-on-the-rise
10. www.guardian.co.uk/lifeandstyle/2011/nov/13/paper-wine-bottle-greenbottle?newsfeed=true
11. www.parducci.com/Parducci-Green/Carbon-Neutral
12. www.reading.ac.uk/food/about/news/foodbio-news-157.aspx
13. www.theglobeandmail.com/life/food-and-wine/wine/beppi-crosariol/ol-fashioned-wine-corks-make-a-comeback/article2181782/
14. http://news.bbc.co.uk/2/hi/business/6270132.stm
15. www.federdoc.com/articoli_notizie/articolo40.html
16. www.cbrands.com/corporate-social-responsibility/sustainability
17. www.popsci.com/technology/article/2010-01/thieves-use-google-earth-evil-plunder-winery-solar-panels
18. www.getsolar.com/News/Massachusetts/Boston-Solar/MIT-Professors-Use-Virus-to-Boost-Solar-Cell-Efficiency---800493435
19. www.stormfisher.com/
20. www.bodegasregalia.es/comunicacion/wp-content/uploads/2011/02/VWM-GRAHAM-GEOTHERM-2.pdf
21. www.bbc.co.uk/science/horizon/2002/eldorado.shtml
22. www.swissinfo.ch/eng/specials/climate_change/research/The_climate_farmer_who_grows_a_mean_pinot.html?cid=33416
23. http://biocharfarms.org/about_biochar/

Further Information

The Wine & Spirit Education Trust

Often referred to as WSET, The Wine & Spirit Education Trust is a British organisation which arranges courses and exams in the field of wine and spirits. WSET was founded in 1969.

www.wsetglobal.com

Wine America

The mission of WineAmerica is to encourage the dynamic growth and development of American wineries and winegrowing through the advancement and advocacy of sound public policy.

www.wineamerica.org

Agence Bio

French agency for the development & promotion of Organic agriculture.

www.agencebio.org

Demeter

Demeter is the brand for products from Biodynamic Agriculture. Only strictly controlled and contractually bound partners are permitted to use the Brand.

www.demeter.net

The National Wine Centre of Australia

The National Wine Centre of Australia is a public exhibition building about winemaking and its industry in South Australia, opened in 2001.It contains an interactive permanent exhibition of winemaking, introducing visitors to the technology, varieties and styles of wine.

www.wineaustralia.com.au

OIV

Created in 1924 by six producer states in response to the global viticultural crisis, the OIV has developed and adapted to become, since 2001, the scientific and technical reference organisation for the entire vitivinicultural field. Its 46 Member States account for more than 85% of global wine production and nearly 80% of world consumption.

www.oiv.int

Institute of Masters of Wine

Through its Members and activities, the Institute of Masters of Wine promotes excellence, interaction and learning, across all sectors of the global wine community.

www.mastersofwine.org

Bibliography

Anstie, Francis, On the Uses of Wine

Beare, Sally, 50 Secrets of the World's Longest Living People

Burk, Kathleen & Bywater, Michael, Is This Bottle Corked? The Secret Life of Wine, Faber and Faber Ltd, London 2008

Bushell, William C., Longevity, Regeneration, and Optimal Health: Integrating ..., Volume 1172: Integrating Eastern and Western Perspectives, 2009

California Wine Advisory Board, Uses of Wine in Medical Practice: A Summary

Chartier, François, Taste Buds and Molecules: The Art and Science of Food with Wine, 2011

Chèze, Cathérine; Vercauteren Joseph and Verpoorte, R., Polyphenols, Wine, and Health: Proceedings of the Phytochemical Society of Europe, Bordeaux, France, 14th-16th April, 1999;

Contreras, Francisco, The Coming Cancer Care, Authentic Lifestyle, UK 2003

Corder, Roger, The Red Wine Diet

Das, Dipak K. and Ursini, Fulvio, Alcohol and Wine in Health and Disease

Doléris, Jacques Amédée, Le Vin Et Les Médecins: Le Pour Et Le Contre

Faith, Nicholas, The Winemasters of Bordeaux, Carlton Books Limited, London 2005

Flynn Siler, Julia, The House of Mondavi, Penguin Group, New York 2007

Frankenburg, Frances Rachel, Vitamin Discoveries and Disasters: History, Science, and Controversies, 2009

Gelb, Michael J., Wine Drinking For Inspired Thinking, Running Press Book Publishers, Philadelphia 2010

Gibson, Michael, The Sommelier Prep Course: An Introduction To the Wines, Beers, and Spirits, 2010

Goldfinger, Tedd and Nicholson, Lynn F., The Wine Lover's Healthy Weight Loss Plan, 2006

Hall, Colin Michael and Mitchell, Richard, Wine Marketing: A Practical Guide

Jackson, Ronald S., Wine Science: Principles and Applications; 2008

Jackson, Ronald S., Wine Science: Principles, Practice, Perception, 2000

Joly, Nicolas, What Is Biodynamic Wine?, Clairview Books, East Sussex 2007

Jukes, Matthew, The Wine Book, Headline Publishing Group, London 2007

Köhnlechner, Manfred, Healing Wines: Celebrating their Curative Powers, 1982

Lucia, Salvatore Pablo, Wine and Health: Proceedings of the First International Symposium On Wine and Health Held At the University of Chicago Center For Continuing Education, November 9, 1968;

Maroon, Joseph and Baur, Joseph, The Longevity Factor: How Resveratrol and Red Wine Activate Genes for a Longer and Healthier Life

Mccarthy, Ed and Ewing-Mulligan, Mary, Wine for Dummies, 2006

Moulton, Kirby S. and Lapsley, James T., Successful Wine Marketing, 2001

O'byrne, Paul, Red Wine and Health

Peck, Garrett, The Prohibition Hangover: Alcohol in America from Demon Rum to Cult Cabernet, 2009

Perdue, Lewis, The Wrath of Grapes: The Coming Wine Industry Shakeout and How to Take Advantage of It

Quinlan Forde, Ralph, The Book of Tibetan Medicine, Octopus Publishing Group Ltd, London 2008

Reader's Digest Guide to Vitamins, Minerals and Supplements, The Reader's Digest Association Limited, London 2000

Reid, Daniel, Guarding the Three Treasures, Simon & Schuster Ltd, London 1993

Resnick, Évelyne, Wine Brands: Success Strategies for New Markets, New Consumers and New Trends, 2008

Robinson, Jancis, Jancins Robinson's Wine Course, BBC Worldwide Ltd, London 1995

Santini, Marco, Wine Country Europe, Rizzoli International Publications, New York 2005

Schäffer, Andrea, Red Wine for Your Health

Schlosser, Eric, Fast Food Nation, Penguin Books, UK 2001

Society of Medical Friends of Wine, Wine Institute (San Francisco, Calif.), Wine, Health and Society, University of California, San Francisco, 1981

Soetaert, Wim and Vandamme, Erick J., Industrial Biotechnology: Sustainable Growth and Economic Success

St Aubyn, Marjorie The Wine Lady, Stay Healthy With Wine

Stanbury, P.F. and Whitaker, A., Principles of Fermentation Technology, Pergamon Press Ltd, England 1987

Sterling, Richard, The Adventure of Food: True Stories of Eating Everything

Taber, George M., Judgement of Paris, Scribner, New York 2005

Tompkins, Peter and Bird, Christopher, The Secret Life of Plants, Harper & Row Publishers Inc., New York 1989

Vogel, Alfred, The Nature Doctor, Mainstream Publishing Company Ltd, Edinburgh 2003

Wallace, Benjamin, The Billionaire's Vinegar, Three River Press, New York 2009

Watkins, Tom R., Wine: Nutritional and Therapeutic Benefits, American Chemical Society. Division of Agricultural and Food Chemistry, American Chemical Society. Meeting, 1997

Acknowledgements

Firstly I would like to thank all the amazing innovators in the wine industry who have inspired me over the months that I have been writing this book. Due to you the world of wine truly deserves a more deeper investigation other than the vineyard name and region. Isco designed the book cover half way through writing the book and gave me the focus to finish the text. Without your cover Isco this would have taken a lot longer. The publisher who released me to do as I please. Dr Maury for his inspiring book which started my journey into popular wine science. Prayot for your patience as I wrote the book. Thanks. The red head for the use of her dining table. Apil for a website that I can manage and that will inform people of the health benefits of wine the world over. He made the site in record time which leaves Kerry internet makers way behind. Suzanne for removing any blocks and proof reading the text. She also did the complicated Kindle formatting and upload. You the reader for taking an interest in the book. Josecho for saying this could be good for me. Javier for his amazing timing for texts.

To anybody and everybody who helped to make the book a success.

To the vine itself for allowing me to tell part of its wonderful story.

The Author

Wine enthusiast Ralph Quinlan Forde is the author of NutriWine and a Holistic Medicine Consultant. He became an award-winning entrepreneur for a complimentary medicine company he set up in Ireland. His first book The Book of Tibetan Medicine is now in 11 language editions world-wide. The book received many positive endorsements from major lights in the alternative and naturopathic medicine field. Being a herbalist and aromatherapist with a knowledge based on number of different complimentary medicines including Tibetan Medicine, Chinese Medicine, Homeopathy and nutrition. He has spoken frequently on BBC radio and national television on these subjects as well as on business and entrepreneurialism. Ralph graduated from The University of Reading with an Honours degree in Biotechnology.

Ralph has contributed to The Irish Examiner, The Sunday Tribune, The Independent On Sunday, IVENUS, FeelGood, Tescos and Health and Fitness magazine. Ralph was shortlisted for the Potter's Health writer of the Year. He has been interviewed by numerous magazines and newspapers for his views on natural health and wellbeing.

He hopes that NutriWine inspires more people to become wine enthusiasts, especially those suffering with heart disease, diabetes and cancer. The book also aims to enrol people into wine culture via exciting innovations and environmental developments taking place in the vineyard.

Climate change is having a huge impact in the world of wine and we have to act now if we still want to be drinking wine in 20 years time.

For further information visit www.nutriwine.net.